Statistics and Computing

Series Editors:
J. Chambers
D. Hand
W. Härdle

Statistics and Computing

Brusco/Stahl: Branch-and-Bound Applications in Combinatorial Data Analysis.
Dalgaard: Introductory Statistics with R.
Gentle: Elements of Computational Statistics.
Gentle: Numerical Linear Algebra for Applications in Statistics.
Gentle: Random Number Generation and Monte Carlo Methods, 2nd Edition.
Härdle/Klinke/Turlach: XploRe: An Interactive Statistical Computing Environment.
Krause/Olson: The Basics of S-PLUS, 4th Edition.
Lange: Numerical Analysis for Statisticians.
Lemmon/Schafer: Developing Statistical Software in Fortran 95
Loader: Local Regression and Likelihood.
Ó Ruanaidh/Fitzgerald: Numerical Bayesian Methods Applied to Signal Processing.
Pannatier: VARIOWIN: Software for Spatial Data Analysis in 2D.
Pinheiro/Bates: Mixed-Effects Models in S and S-PLUS.
Venables/Ripley: Modern Applied Statistics with S, 4th Edition.
Venables/Ripley: S Programming.
Wilkinson: The Grammar of Graphics.

Michael J. Brusco
Stephanie Stahl

Branch-and-Bound Applications in Combinatorial Data Analysis

Springer

Michael J. Brusco
Department of Marketing
College of Business
Florida State University
Tallahassee, FL 32306-1110
USA

Stephanie Stahl
2352 Hampshire Way
Tallahassee, FL 32309-3138
USA

Series Editors:
J. Chambers
Bell Labs, Lucent Technologies
600 Mountain Avenue
Murray Hill, NJ 07974
USA

D. Hand
Department of Mathematics
South Kensington Campus
Imperial College, London
London SW7 2AZ
United Kingdom

W. Härdle
Institut für Statistik und Ökonometrie
Humboldt-Universität zu Berlin
Spandauer Str. 1
D-10178 Berlin
Germany

Library of Congress Control Number: 2005924426

ISBN-10: 0-387-25037-9   Printed on acid-free paper.
ISBN-13: 978-0387-25037-3

© 2005 Springer Science+Business Media, Inc.
All rights reserved. This work may not be translated or copied in whole or in part without the written permission of the publisher (Springer Science+Business Media, Inc., 233 Spring Street, New York, NY 10013, USA), except for brief excerpts in connection with reviews or scholarly analysis. Use in connection with any form of information storage and retrieval, electronic adaptation, computer software, or by similar or dissimilar methodology now known or hereafter developed is forbidden.
The use in this publication of trade names, trademarks, service marks, and similar terms, even if they are not identified as such, is not to be taken as an expression of opinion as to whether or not they are subject to proprietary rights.

Printed in the United States of America.   (EB)

9 8 7 6 5 4 3 2 1

springeronline.com

For Cobol Lipshitz and Snobol Gentiment

Preface

This monograph focuses on the application of the solution strategy known as *branch-and-bound* to problems of *combinatorial data analysis*. Combinatorial data analysis problems typically require either the selection of a subset of objects from a larger (master) set, the grouping of a collection of objects into mutually exclusive and exhaustive subsets, or the sequencing of objects. To obtain verifiably optimal solutions for this class of problems, we must evaluate (either explicitly or implicitly) all feasible solutions. Unfortunately, the number of feasible solutions for problems of combinatorial data analysis grows exponentially with problem size. For this reason, the explicit enumeration and evaluation of all solutions is computationally infeasible for all but the smallest problems. The branch-and-bound solution method is one type of *partial enumeration* solution strategy that enables some combinatorial data analysis problems to be solved optimally without explicitly enumerating all feasible solutions.

To understand the operation of a branch-and-bound algorithm, we distinguish complete solutions from partial solutions. A complete solution is one for which a feasible solution to the optimization problem has been produced (e.g., all objects are assigned to a group, or all objects are assigned a sequence position). A partial solution is an incomplete solution (e.g., some objects are not assigned to a group, or some objects are not assigned a sequence position). During the execution of a branch-and-bound algorithm, solutions are gradually constructed and are, therefore, only partially completed at most stages. If we can determine that a partial solution cannot possibly lead to an optimal solution, then that partial solution and all possible complete solutions stemming from the partial solution can be eliminated from further consideration. This elimination of partial solutions and, simultaneously, all of the complete and partial solutions that could be generated from them is the cornerstone of the branch-and-bound method. In the worst case, a branch-and-bound algorithm could require the complete enumeration of all feasible solutions. For this reason, we note from the outset that we will reach a point where branch-and-bound is not computationally feasible, and other solution approaches are required.

We develop and describe a variety of applications of the branch-and-bound paradigm to combinatorial data analysis. Part I of the monograph (Chapters 2 through 6) focuses on applications for partitioning a set of objects based on various criteria. Part II (Chapters 7 through 11) describes branch-and-bound approaches for seriation of a collection of objects. Part III (Chapters 12 through 14) addresses the plausibility of branch-and-bound methods for variable selection in multivariate data analysis, particularly focusing on cluster analysis and regression.

Our development of this monograph was largely inspired by the recent monograph by Hubert, Arabie, and Meulman (2001) titled, *Combinatorial Data Analysis: Optimization by Dynamic Programming*. Like the branch-and-bound method, dynamic programming is a partial enumeration strategy that can produce optimal solutions for certain classes of combinatorial data analysis problems that would be insurmountable for exhaustive enumeration approaches. Many of the problems tackled in Parts I and II are also addressed in the Hubert et al. (2001) monograph, which would make an excellent companion reading for our monograph.

At the end of many of the chapters, we identify available computer programs for implementing the relevant branch-and-bound procedures. This software is offered (without warranty) free of charge, and both the source code and executable programs are available. This code should enable users to reproduce results reported in the chapters, and also to implement the programs for their own data sets.

We are deeply indebted to Phipps Arabie, J. Douglas Carroll, and Lawrence Hubert, whose enthusiasm and encouragement were the principal motivating factors for our pursuing this area of research. We also thank J. Dennis Cradit, who has frequently collaborated with us on related problems of combinatorial data analysis.

Michael J. Brusco
Stephanie Stahl

January, 2005

Contents

Preface vii

1 Introduction 1
   1.1 Background 1
   1.2 Branch-and-Bound 4
      1.2.1 A Brief History 4
      1.2.2 Components of a Branch-and-Bound Model 6
   1.3 An Outline of the Monograph 9
      1.3.1 Module 1: Cluster Analysis–Partitioning 9
      1.3.2 Module 2: Seriation 10
      1.3.3 Module 3: Variable selection 11
   1.4 Layout for Nonintroductory Chapters 11

I   Cluster Analysis—Partitioning   13

2 An Introduction to Branch-and-Bound Methods for Partitioning 15
   2.1 Partitioning Indices 16
   2.2 A Branch-and-Bound Paradigm for Partitioning 20
      2.2.1 Algorithm Notation 20
      2.2.2 Steps of the Algorithm 21
      2.2.3 Algorithm Description 21

3 Minimum-Diameter Partitioning 25
   3.1 Overview 25
   3.2 The INITIALIZE Step 26
   3.3 The PARTIAL SOLUTION EVALUATION Step 30
   3.4 A Numerical Example 32
   3.5 Application to a Larger Data Set 34
   3.6 An Alternative Diameter Criterion 38
   3.7 Strengths and Limitations 39
   3.8 Available Software 39

4 Minimum Within-Cluster Sums of Dissimilarities Partitioning — 43
4.1 Overview 43
4.2 The INITIALIZE Step 44
4.3 The PARTIAL SOLUTION EVALUATION Step 46
4.4 A Numerical Example 50
4.5 Application to a Larger Data Set 53
4.6 Strengths and Limitations of the Within-Cluster Sums Criterion 54
4.7 Available Software 56

5 Minimum Within-Cluster Sums of Squares Partitioning — 59
5.1 The Relevance of Criterion (2.3) 59
5.2 The INITIALIZE Step 60
5.3 The PARTIAL SOLUTION EVALUATION Step 64
5.4 A Numerical Example 66
5.5 Application to a Larger Data Set 70
5.6 Strengths and Limitations of the Standardized Within-Cluster Sum of Dissimilarities Criterion 71
5.7. Available Software 73

6 Multiobjective Partitioning — 77
6.1 Multiobjective Problems in Cluster Analysis 77
6.2 Partitioning of an Object Set Using Multiple Bases 77
6.3 Partitioning of Objects in a Single Data Set Using Multiple Criteria 82
6.4 Strengths and Limitations 84
6.5 Available Software 85

II Seriation — 89

7 Introduction to the Branch-and-Bound Paradigm for Seriation — 91
7.1 Background 91
7.2 A General Branch-and-Bound Paradigm for Seriation 93

8 Seriation—Maximization of a Dominance Index — 97
8.1 Introduction to the Dominance Index 97
8.2 Fathoming Tests for Optimizing the Dominance Index 98
    8.2.1 Determining an Initial Lower Bound 98
    8.2.2 The Adjacency Test 100
    8.2.3 The Bound Test 100
8.3 Demonstrating the Iterative Process 102
8.4 EXAMPLES—Extracting and Ordering a Subset 103

    8.4.1 Tournament Matrices 103
    8.4.2 Maximum Dominance Index vs. Perfect Dominance for
         Subsets 106
8.5 Strengths and Limitations 110
8.6 Available Software 111

9 Seriation—Maximization of Gradient Indices   113
9.1 Introduction to the Gradient Indices 113
9.2 Fathoming Tests for Optimizing Gradient Indices 115
    9.2.1 The Initial Lower Bounds for Gradient Indices 115
    9.2.2 The Adjacency Test for Gradient Indices 116
    9.2.3 The Bound Test for Gradient Indices 121
9.3 EXAMPLE—An Archaeological Exploration 123
9.4 Strengths and Limitations 124
9.5 Available Software 127

10 Seriation—Unidimensional Scaling   129
10.1 Introduction to Unidimensional Scaling 129
10.2 Fathoming Tests for Optimal Unidimensional Scaling 131
    10.2.1 Determining an Initial Lower Bound 131
    10.2.2 Testing for Symmetry in Optimal Unidimensional
          Scaling 131
    10.2.3 Adjacency Test Expanded to Interchange Test 132
    10.2.4 Bound Test 136
10.3 Demonstrating the Iterative Process 139
10.4 EXAMPLE—Can You Hear Me Now? 141
10.5 Strengths and Limitations 144
10.6 Available Software 144

11 Seriation—Multiobjective Seriation   147
11.1 Introduction to Multiobjective Seriation 147
11.2 Efficient Solutions 149
11.3 Maximizing the Dominance Index for Multiple
    Asymmetric Proximity Matrices 149
11.4 UDS for Multiple Symmetric Dissimilarity Matrices 153
11.5 Comparing Gradient Indices for a Symmetric
    Dissimilarity Matrix 160
11.6 Multiple Matrices with Multiple Criteria 164
11.7 Strengths and Limitations 169

III   Variable Selection   171

12 Introduction to Branch-and-Bound Methods for Variable Selection   173
    12.1 Background 173

13 Variable Selection for Cluster Analysis   177
    13.1 *True* Variables and *Masking* Variables 177
    13.2 A Branch-and-Bound Approach to Variable Selection 178
    13.3 A Numerical Example 182
    13.4. Strengths, Limitations, and Extensions 183

14 Variable Selection for Regression Analysis   187
    14.1 Regression Analysis 187
    14.2 A Numerical Example 190
    14.3 Application to a Larger Data Set 193
    14.4 Strengths, Limitations, and Extensions 198
    14.5 Available Software 199

Appendix A: General Branch-and-Bound Algorithm for Partitioning   203

Appendix B: General Branch-and-Bound Algorithm Using Forward Branching for Optimal Seriation Procedures   205

References   209

Index   219

1 Introduction

1.1 Background

There are many problems in statistics that require the solution of continuous (or smooth) optimization problems. Examples include maximum-likelihood estimation for confirmatory factor analysis and gradient-based search methods for multidimensional scaling. These problems are often characterized by establishment of an objective function and, in some instances, a set of corresponding constraints. Solution proceeds via standard calculus-based methods for (constrained or unconstrained) optimization problems. In some situations, the optimality conditions associated with the calculus-based approach enable a closed form solution to be obtained. An example is the classic normal equation for linear regression analysis. In other circumstances, no closed form solution exists and the problem must be solved via numerical estimation procedures, such as the Newton-Raphson method.

Although they are somewhat less emphasized in the statistical literature, a variety of data analysis scenarios require the solution of *discrete optimization problems*. Unlike continuous optimization problems, discrete optimization problems are not smooth functions because integer restrictions are placed on (at least some of) the relevant variables. For discrete problems, no set of optimality conditions can be easily checked to guarantee optimality. Instead, demonstration of optimality for a discrete optimization problem typically requires evaluation, either explicitly or implicitly, of the complete set of feasible solutions to the problem. Discrete optimization problems are, therefore, solved by enumerative methods that investigate the feasible solution set. The particular types of discrete optimization problems of interest in this monograph fall under the area of *combinatorial data analysis*. The word *combinatorial* implies that the problem of interest involves "combinatorial choices," such as ordering of objects, grouping a collection of objects, or selecting a subset of a larger set of objects.

To conceptualize the nature of a combinatorial optimization problem, we will begin with an example of combinatorial optimization that is not explicitly related to data analysis, yet has an intuitive conceptual appeal. Consider a plant manager who wishes to sequence $n = 6$ workstations along a manufacturing line. Without loss of generality, we can assume that the flow along the line is left to right. The manager has information concerning the daily workflow from each station to every other station, which is provided in the form of a six-by-six matrix, $\mathbf{W} = \{w_{ij}\}$, where w_{ij} is the daily flow from workstation i to workstation j (for $1 \leq i \neq j \leq 6$). Some sample workflow data are shown in Table 1.1. Backtracking of workflow corresponds to workflow moving from right to left (backwards) along the line. The manager would like to avoid backtracking to the greatest extent possible, and is faced with the following combinatorial optimization problem:

- *Minimum-backtracking row layout problem.* Find a sequence of the six workstations that will minimize the total amount of daily backtracking.

Table 1.1. A 6×6 workflow matrix, $\mathbf{W}$, for a minimum backtracking sequencing example.

		Flow to workstation					
		1	2	3	4	5	6
Flow from workstation	1	---	0	10	40	0	10
	2	30	---	10	10	20	0
	3	0	5	---	0	10	30
	4	5	0	5	---	10	0
	5	0	10	20	50	---	20
	6	0	0	5	20	0	---

The plant manager could find the optimal solution to this problem by enumerating all feasible sequences and selecting the one that corresponded to the least amount of backtracking. To determine the number of possible sequences, the manager determines that there are six possible assignments for the leftmost position in the sequence. Once the leftmost position is assigned a workstation, there are five possibilities for the second leftmost position, and so on. Thus, the feasible set of solutions for this problem consists of $n! = 6 \times 5 \times 4 \times 3 \times 2 \times 1 = 720$ possible sequences. The best of these sequences is 2-1-5-3-6-4, which yields total backtracking of 50, as corresponding to lower triangle of the reordered matrix shown in Table 1.2.

Table 1.2. Reordered workflow matrix, **W**, for the optimal sequence 2-1-5-3-6-4.

		Flow to workstation					
		2	1	5	3	6	4
Flow from	2	---	30	20	10	0	10
workstation	1	0	---	0	10	10	40
	5	10	0	---	20	20	50
	3	5	0	10	---	30	0
	6	0	0	0	5	---	20
	4	0	5	10	5	0	---

Although exhaustive enumeration of the feasible solution set was computationally viable for this example, the plant manager would have been unable to implement such an approach if there were $n = 16$ workstations instead of only six. The total number of feasible sequences for 16 workstations is 16!, which is more than twenty trillion. Fortunately, there are a variety of *partial enumeration strategies* that would facilitate solution of the 16-workstation problem.

Partial enumeration methods can often provide optimal solutions to combinatorial optimization problems without the need for explicit enumeration of the entire feasible solution set. We note, however, at this early juncture of the monograph, that partial enumeration methods are still prone to serious limitations on the sizes of problems they can handle. They do not provide a "quick fix" to the combinatorial nature of our optimization problems but they do often significantly expand the size of problems for which optimal solutions can be obtained.

There are two principal partial enumeration strategies in the combinatorial optimization literature: (a) dynamic programming, and (b) branch-and-bound. In a recent monograph, Hubert, Arabie, and Meulman (2001) eloquently outlined a dynamic programming paradigm that is broadly applicable to a variety of problems in combinatorial data analysis. In particular, they demonstrated that this paradigm could be used to solve a host of problems related to cluster analysis and seriation. They have also made a Fortran library available for the applications they described. As the authors note, the primary limiting factor of the dynamic programming approach for the described applications is its sensitivity to computer storage capacity.

Branch-and-bound is also an attractive partial enumeration strategy for optimization problems in combinatorial data analysis; however, there is currently no analogous resource to Hubert et al.'s (2001) dynamic programming monograph. Our goal in this monograph is to fill this void. As we shall see, the branch-and-bound approach offers some advantages over dynamic programming for certain classes of problems, but also has some drawbacks that must be carefully addressed for successful implementation. The branch-and-bound algorithms described in this monograph require far less computer storage than their dynamic programming competitors, which makes the former class of algorithms more amenable to handling larger problems. On the other hand, dynamic programming methods are much less sensitive to the characteristics of the input data and are, therefore, more consistent with respect to solution times for problems of a given size. Branch-and-bound algorithms are quite sensitive to the properties of the data. For some data sets, branch-and-bound might be more efficient than dynamic programming, whereas in other cases branch-and-bound can require much more computation time than dynamic programming. The key is to recognize that neither approach is superior in all cases and consider the data properties when choosing a methodology.

1.2 Branch-and-Bound

1.2.1 A Brief History

According to Murty (1995, Chapter 10), the branch-and-bound approach to optimization problems was developed independently by Land and Doig (1960) and Murty, Karel, and Little (1962) [see also Little, Murty, Sweeney, and Karel (1963)]. One possible way of clarifying this simultaneous development is to recognize that Land and Doig (1960) were focusing on general discrete programming problems (or mixed discrete programming problems, which contain some discrete variables and some continuous variables), whereas Murty et al. (1962) focused on a specific type of discrete optimization problem.

Land and Doig's (1960) branch-and-bound method was developed within the context of solving linear programs with integer decision variables. When attempting to solve an integer programming model using standard linear programming techniques (e.g., the simplex method), one or more of the variables that are required to assume integer values might be fractional in the actual linear programming solution. The branching process in Land and Doig's algorithm branches from these fractional variables. One branch requires the fractional variable to assume a value less than or equal to the largest integer that is less than the fractional value, whereas the other branch requires the variable to assume a value that is greater than or equal to the smallest integer that is greater than the fractional value. The algorithm continues with the solution of linear programming subproblems for the nodes of the branches. The ultimate goal is to produce an optimal solution to the linear programming problem that satisfies the integer restrictions on the appropriate variables.

Branch-and-bound implementation in our monograph is much closer to the method for the traveling salesman problem as approached by Murty et al. (1962) and Little et al. (1963). The traveling salesman problem is a classic combinatorial optimization problem that can be conceptually described as follows:

- A salesperson must leave a city, visit a set of $n - 1$ other cities, and return to the city of origin so as to minimize travel distance.

The traveling salesman problem obviously has applications for problems such as the routing of personnel or vehicles, but also has many interesting applications in combinatorial data analysis (see, for example, Hubert and Baker [1978]). There are other discrete optimization problems that also have particular relevance for data analysis. These problems include, but are not limited to (a) quadratic assignment, (b) 0-1 knapsack, (c) set partitioning/covering, (d) bin-packing, and (e) *p*-median. The *quadratic assignment problem* (Koopmans & Beckmann, 1957), which actually subsumes the traveling salesman problem as a special case, is well recognized for its usefulness in the analysis of statistical data (Hubert, 1987, Chapter 4; Hubert & Schultz, 1976). The *knapsack problem* has particular applicability to selection of subsets of objects from a larger set (Brusco & Stahl, 2001b). The *set partitioning problem*, as the term suggests, has a natural relationship to the partitioning of a data set. Set covering relaxes the need for a partition and, therefore, might lend itself to the case of overlapping clusters. The *p*-median problem is a well-studied integer programming problem that also has direct relevance to cluster analysis (Klastorin, 1985; Mulvey & Crowder, 1979).

We do not contend that branch-and-bound is the best optimal solution procedure for the discrete optimization problems mentioned in the preceding paragraph. Indeed, the most effective optimal solution procedures for these problems are principally based on or facilitated by other techniques, such as Lagrangean relaxation, column generation, or cutting planes. Accordingly, our focus will not be on the discussion of optimal methods for these classic discrete optimization problems, for which there are several excellent sources available (Murty, 1995; Parker & Rardin, 1988). Instead, our emphasis is on specific applications of the branch-and-bound method to problems of data analysis, for which a single, integrative resource does not currently exist.

1.2.2 Components of a Branch-and-Bound Model

The branch-and-bound algorithms implemented in our monograph consist of a variety of components, some of which vary based on the particular application. Nevertheless, there are several major components that are relevant to all of the algorithms: (a) *branching*, (b) *bounding*, (c) *pruning*, and (d) *retracting*. A branching step corresponds to the creation of one or more new subproblems from an existing subproblem. For example, returning to the minimum backtracking layout example in Table 1.1, suppose that a *partial sequence* is available that corresponds to workstations 2 and 4 being allocated to the first two positions in the sequence, respectively. We can create a branch from this partial solution by assigning workstation 1 to the third position of the sequence. Throughout the monograph, we will refer to this type of step as a *branch forward* because the partial solution is "moving forward" by assigning another object. In other words, if we view the branching process in terms of levels, the assignment of two workstations would correspond to the second level of the process. The assignment of a third workstation moves us to the third level.

We will also refer to *branch right* operations, which correspond to a change in the partial solution while remaining at the same level. For example, if we changed the workstation in position 2 from 4 to 5, then we are remaining at the second level despite the fact that we are now pursuing another branch. This is an example of a branch right operation. Branch right operations are typically preceded by either a pruning operation or retraction, which are described below.

Bounding is a crucial aspect of the algorithms described in this monograph, and there are several aspects to this process. One important ingredient of the bounding process is that it will generally be assumed for minimization (maximization) problems that an initial *upper* (*lower*) *bound* on the optimal objective criterion value is available. In most cases, a heuristic method can be efficiently applied to establish an initial bound. For example, in the workstation sequencing example, an initial upper bound might be provided by the manager's trial and error attempts to sequence the stations. Assuming that this process produced the sequence 2-5-1-6-4-3, the total backtracking associated with this sequence is 75, and this figure could be supplied as an upper bound for a branch-and-bound algorithm.

The upper (lower) bound is used to evaluate partial solutions during the implementation of the branch-and-bound algorithm. A partial solution generally requires the consideration of two pieces of information that can be combined to provide a value that represents a limit on the best possible objective criterion value that could be realized from completion of the partial sequence. The first piece of information corresponds to a direct contribution to the objective function from the partial solution itself. The second source of information corresponds to an estimate of the most favorable contribution to the objective function that could be realized from completing the partial solution.

Upon the establishment of the best possible solution that could possibly be realized for a partial solution, a comparison is made to the current upper (lower) bound. If the current partial solution falls short of the current upper bound for minimization problem (or exceeds the current lower bound for a maximization problem), then the current partial solution cannot possibly lead to a solution that is better than the incumbent solution. At this point, a *pruning* operation is implemented. In other words, the current partial solution cannot possibly lead to an optimal solution and, thus, the current solution and all possible branches and completed solutions that could stem from it are eliminated from further consideration. This is a fundamental aspect of the branch-and-bound algorithm, the earlier a partial solution is pruned, the greater the number of complete solutions that will need to be explicitly evaluated.

As its name implies, *retraction* corresponds to "moving backward" in the partial solution. In other words, retraction involves moving backward from a higher level to a lower level. For example, suppose the current partial sequence for the workstation sequencing problem is 2-4-6, and a branch right operation is encountered. This would produce the sequence 2-4-7, which obviously does not make sense because there are only six workstations. Instead, the correct operation at this point is retraction, which would begin by removing any assignment to the third position and, subsequently stepping back to the second position (i.e., moving from level three to level two). Next, a branch right operation would be applied to the second position, resulting in the new partial sequence 2-5. The implication is that all possible solutions from the partial sequence 2-4 had been either implicitly or explicitly evaluated, and the next step was to move back to the second level and build branches from 2-5.

To illustrate the components of branch-and-bound, we return one final time to the minimum backtracking example. We assume the initial upper bound is 75 and that the current partial sequence is 1-5. The total backtracking that corresponds to this partial sequence is $(w_{21} + w_{31} + w_{41} + w_{51} + w_{61} + w_{25} + w_{35} + w_{45} + w_{65}) = (30 + 0 + 5 + 0 + 0 + 20 + 10 + 10 + 0) = 75$. We know that any completed sequence must have a backtracking total of at least 75 because we know that workstations 1 and 5 are the first two in the sequence. However, we can augment this term by recognizing that the very best possible contributions corresponding to the yet unassigned workstations are given by $\min(w_{23} + w_{32}) + \min(w_{24} + w_{42}) + \min(w_{26} + w_{62}) + \min(w_{34} + w_{43}) + \min(w_{36} + w_{63}) + \min(w_{46} + w_{64}) = 5 + 0 + 0 + 0 + 5 + 0 = 10$. Adding the two components together yields $75 + 10 = 85$, which is greater than the current upper bound of 75. Thus, the partial sequence of two workstations is *pruned*, and all $(6-2)! = 24$ possible sequences that could stem from this partial sequence need not be explicitly evaluated.

After pruning the partial sequence 1-5, we would *branch right* to the partial sequence 1-6. The direct contribution to backtracking from this partial sequence is 85, which exceeds the upper bound. Thus, the partial sequence would be pruned. Branching right is not appropriate here, as there is not a seventh workstation. Therefore, *retraction* occurs by moving back to level one and branching right, thus creating a new partial solution that simply consists of workstation 2 in the first position.

When considering the partial sequence consisting of workstation 2 in the first position, the direct contribution to backtracking is $(w_{12} + w_{32} + w_{42} + w_{52} + w_{62}) = (0 + 5 + 0 + 10 + 0) = 15$. Adding the minimum possible contributions that could possibly occur from pairs of the unassigned workstations yields a value of 30. Because $15 + 30 = 45 < 75$, the branch-and-bound algorithm would proceed by branching forward (i.e., moving from the first level to the second level) and creating the new partial sequence 2-1. Pursuing this branch would ultimately lead to the optimal sequence 2-1-5-3-6-4, which was noted as the optimal solution in section 1.1.

1.3 An Outline of the Monograph

Throughout the remainder of this monograph, we will focus on important application areas of the branch-and-bound method in combinatorial data analysis. The heaviest focus will be placed on applications to partitioning models in cluster analysis and seriation, which are presented in Chapters 2 through 11. These chapters will serve as an excellent parallel to Hubert et al.'s (2001) dynamic programming monograph, which focuses principally on cluster analysis and seriation issues. The final three chapters are devoted to branch-and-bound methods for variable (or feature) selection problems. Variable selection issues are broadly applicable to a number of problems in multivariate statistics, including cluster analysis, discriminant analysis, and regression.

1.3.1 Part I: Cluster Analysis–Partitioning

Cluster analysis is an important multivariate statistical technique that spans areas of interest across both the physical and behavioral sciences. In Chapter 2, we introduce a general branch-and-bound paradigm that can be used to produce optimal clustering solutions to problems of nontrivial size. Chapters 3, 4, and 5 build on this paradigm to illustrate specific modeling issues and demonstrations for several important criteria. Chapter 3 focuses on minimum diameter partitioning, which is a well-known criterion with close links to the important problem of graph coloring. Chapter 4 emphasizes the minimization of within-cluster sums of dissimilarity elements, and Chapter 5 considers a similar criterion but standardizes the dissimilarity sums by the number of objects within the clusters. An important aspect of Chapter 5 is its relationship to minimization of a within-cluster sum of squared error, perhaps the most frequently employed criterion in the cluster analysis literature.

The cluster analysis part of the monograph concludes with a discussion of multiobjective cluster analysis in Chapter 6. This chapter was developed for two reasons. First, modification of the branch-and-bound approach for multiple objective criteria can be rather straightforward. In some cases, considering multiple criteria can increase solution efficiency relative to solution time for a single criterion. Second, consideration of multiple objective criteria in a cluster analysis has several areas of application.

1.3.2 Part II: Seriation

Seriation corresponds to the construction of an order for a collection of objects. Our minimum backtracking workstation sequencing example described in this chapter is an example of seriation. Specifically, we were seriating the workstations to minimize backtracking distance. Seriation problems also have relevance to other problems in operations management, such as the sequencing of jobs to maximize efficiency or optimizing satisfaction of customer due dates. Of course, our emphasis on seriation in this monograph is not on operations management applications, but rather within the context of data analysis. Fortunately, there is no shortage of interest on seriation in the combinatorial data analysis literature, with applications in archaeology, biometrics, and psychometrics. Our material in Part II draws from many of these areas.

As the introductory chapter, Chapter 7 provides the branch-and-bound paradigm for seriation, which is maintained throughout Part II. In Chapter 8 we focus on the reordering of the rows and columns of an asymmetric proximity matrix with the objective of maximizing the sum of the elements above the main diagonal of the reordered matrix. This criterion, which is known as the dominance index, has relevance to applications in paired comparison ranking and social choice. We also address an important variation of this problem, which corresponds to the simultaneous identification of a subset of objects from the complete set and the ordering of the subset.

Chapter 9 is concerned with seriation of symmetric matrices so as to maximize within row and/or column gradient indices. The goal is to produce an ordering that reveals a strong patterning of the elements when moving along the rows or columns. Chapter 10 extends the concept of within row and column gradients to the problem of scaling the objects along a number line. Thus, in this chapter, not only is an ordering of the objects generated but also a distance between each pair of adjacent objects. The goal is to yield a scaling of the objects that minimizes (in a least-squares sense) the difference between the pairwise distances between objects and the corresponding dissimilarity measures.

Chapter 11 provides coverage of multiobjective programming applications in seriation and can be perceived as a counterpart to Chapter 6 in Part I. The chapter covers the seriation of a single matrix based on multiple criteria as well as the seriation of multiple matrices (or multiple sources of information from a single matrix) using one or more criteria.

1.3.3 Part III: Variable Selection

The third and final part of this monograph considers applications of branch-and-bound within the context of variable selection. Chapter 12 provides an introductory overview of various problems and criteria for variable selection. Chapter 13 describes a particular application related to the selection of variables for cluster analysis. An interesting aspect of this particular chapter is that the proposed solution method embeds the cluster analysis branch-and-bound model for the problem posed in Chapter 5 within a branch-and-bound algorithm that seeks to determine the appropriate subset of candidate variables. Not surprisingly, this embedding limits the size of problems that can be tackled, but the principle of embedded algorithms is nonetheless interesting.

Chapter 14 focuses on variable selection within the context of regression analysis. Although our branch-and-bound paradigm for this problem is straightforward in its implementation, perhaps leaving room for a variety of computational enhancements, our principal focus is to demonstrate clearly how branch-and-bound operates for this particular problem.

1.4 Layout for Nonintroductory Chapters

The first chapter of each part of the monograph contains an overview of the particular data analysis problem. For the cluster analysis and seriation parts, a general branch-and-bound paradigm is also provided in the initial chapter of each part. The remaining chapters of each part describe the particular details of algorithmic implementation for the problem under consideration. These details are typically followed by one or more illustrative examples that demonstrate the implementation of the algorithm. The goal is to provide readers with a clear blueprint with respect to how the algorithms actually produce an optimal solution. Several larger examples are also provided to more fully illustrate the methods, and to show that the selection of different objective criteria can lead to rather different solutions. Sections describing limitations of the criterion and/or the application of the branch-and-bound algorithm for the particular problem under consideration are also provided in most chapters.

Although the principal focus of this monograph is not software development, we have made available links to a number of free software programs for the methods described herein. These programs can be accessed and downloaded from the websites of the authors, which are currently maintained at http://psiheart.com/quantpsych/monograph.html (Stahl) and http://garnet.acns.fsu.edu/~mbrusco (Brusco). These websites provide the source codes, a file containing the data sets used in this monograph, and, when applicable, downloadable executable files. In each chapter, the final section describes the available software programs associated with problems in that chapter. These descriptions include the required input format, algorithmic implementation details, and a program output. At least one example of program implementation is typically provided, along with corresponding output. Most of the software programs are written in Fortran, and were complied using Compaq Visual Fortran (version 6.5). All reported results were obtained on a microcomputer with 1 GB of random-access-memory and a 2.2 GHz Intel Pentium® IV processor (Pentium® is a trademark of Intel Corporation).

Part I

Cluster Analysis–Partitioning

2 An Introduction to Branch-and-Bound Methods for Partitioning

We define S as a collection of n objects from which an $n \times n$ symmetric dissimilarity matrix **A** is constructed. For the purposes of cluster analysis, **A** will generally be regarded as having a dissimilarity interpretation, such that larger matrix elements reflect less similarity between the corresponding object pair. The diagonal elements of **A** are irrelevant to all analyses, and off-diagonal elements are assumed to be nonnegative. Throughout most of Part I, the broader definition of "dissimilarity" will be assumed. However, there are portions of Part I that require the more restrictive assumption that the elements of **A** correspond to squared Euclidean distances between pairs of objects. This assumption will be carefully noted when required.

Cluster analysis methods can be classified into a variety of categories. A traditional classification scheme is based on hierarchical versus nonhierarchical (or partitioning) methods. Hierarchical methods produce treelike structures knows as dendrograms, which illustrate hierarchies among the object set. These hierarchical methods are typically further subdivided into two categories: (a) agglomerative algorithms, and (b) divisive methods. Agglomerative hierarchical methods begin with each of the n objects in their own separate cluster, and the dendrogram is produced by iteratively merging clusters until all objects are in the same cluster. Among the most popular hierarchical methods are single-linkage clustering, complete-linkage clustering, and Ward's (1963) minimum variance method. Divisive hierarchical clustering methods begin with all objects in a single cluster and the dendrogram is produced by splitting clusters until each object is in its own individual cluster. Bipartitioning splitting rules, which divide a selected cluster into two clusters at each level of the hierarchy, are especially popular (Guénoche, Hansen, & Jaumard, 1991; Hansen, Jaumard, & Mladenovic, 1998).

Nonhierarchical or partitioning methods produce a separation of objects into K distinct clusters. A partition can be formed by using a hierarchical clustering procedure and cutting the tree at the desired number of clusters. However, such an approach will generally not produce an optimal solution for a particular partitioning problem. The most commonly used partitioning procedures are heuristic algorithms, such as the popular *K*-means algorithm, that efficiently produce good solutions to large-scale problems but do not guarantee that a globally optimal solution will be obtained. Nevertheless, optimal methods for partitioning can be successfully applied for problems that are nontrivial in size. Branch-and-bound algorithms are among the most successful optimal methods for partitioning (Brusco, 2003, 2005; Brusco & Cradit, 2004; Diehr, 1985; Hansen & Delattre, 1978; Klein & Aronson, 1991; Koontz, Narendra, & Fukunaga, 1975).

Throughout the remainder of this Part I, we will focus on branch-and-bound algorithms for partitioning problems. This does not imply that branch-and-bound strategies are not amenable to hierarchical clustering problems, but the preponderance of branch-and-bound related applications in cluster analysis are for partitioning. This chapter continues with a brief overview of some of the most important partitioning indices, and a description of the general branch-and-bound paradigm. Subsequent chapters will present specific algorithmic details and examples for selected criteria.

2.1 Partitioning Indices

There are many possible indices that can be used to develop a partition of the objects in S based on the dissimilarity information in $\mathbf{A}$ (see, for example, Guénoche, 2003; Hansen & Jaumard, 1997; Hubert et al., 2001, Chapter 3). Our development of these indices uses the following notation:

K = the number of clusters, indexed $k = 1,..., K$;

C_k = the set of objects assigned to cluster k ($k = 1,..., K$);

n_k = the number of objects in cluster k, which is the cardinality of C_k for $k = 1,..., K$ (i.e., $n_k = |C_k|$, for $k = 1,..., K$);

π_K = a feasible partition of K clusters, $(C_1, C_2,..., C_K)$;

Π_K = the set of all feasible partitions of size K $\{(\pi_K = \{C_1, C_2,..., C_K\}) \in \Pi_K\}$.

2.1 Partitioning Indices

A feasible partition of n objects into K clusters, π_K, is characterized by the properties of mutually exclusive and exhaustive clusters. More formally, the conditions for a partition are:

Condition i. $C_k \neq \emptyset$ for $1 \leq k \leq K$;

Condition ii. $C_h \cap C_k = \emptyset$ for $1 \leq h < k \leq K$;

Condition iii. $C_1 \cup C_2 \cup ... \cup C_K = S$.

With these definitions in place, some important partitioning criteria can be represented as follows:

$$\min_{\pi_K \in \Pi_K} : f_1(\pi_K) = \max_{k=1,...,K} \left(\max_{(i<j) \in C_k} (a_{ij}) \right), \quad (2.1)$$

$$\min_{\pi_K \in \Pi_K} : f_2(\pi_K) = \sum_{k=1}^{K} \sum_{(i<j) \in C_k} a_{ij}, \quad (2.2)$$

$$\min_{\pi_K \in \Pi_K} : f_3(\pi_K) = \sum_{k=1}^{K} \left(\frac{\sum_{(i<j) \in C_k} a_{ij}}{n_k} \right), \quad (2.3)$$

$$\min_{\pi_K \in \Pi_K} : f_4(\pi_K) = \sum_{k=1}^{K} \left(\frac{\sum_{(i<j) \in C_k} a_{ij}}{n_k (n_k - 1)/2} \right). \quad (2.4)$$

All of these criteria are concerned with the dissimilarity between pairs of objects within the same cluster, known as the dissimilarity elements. For a given cluster k, the maximum dissimilarity element between any pair of objects within that cluster is referred to as the *cluster diameter*. The maximum cluster diameter across all k clusters is termed the *partition diameter*, and this is associated with the objective criterion (2.1). Minimization of the partition diameter produces clusters that are *compact* in the sense that the large dissimilarity elements are suppressed by placing more dissimilar objects in different clusters. Criterion (2.1) is fundamentally different from (2.2), (2.3), and (2.4) in the sense that it minimizes the maximum dissimilarity element among objects in the same clusters, whereas the latter three criteria focus on the sum of the dissimilarity elements within the clusters.

Criterion (2.2) corresponds to the minimization of the sum of the unadjusted within-cluster sums of the dissimilarity measures. Taking this a step further, Criterion (2.3) adjusts the within-cluster sums for cluster size prior to summation across clusters. This criterion is particularly important when **A** contains squared Euclidean distances between pairs of objects. In such cases, criterion (2.3) corresponds to minimization of the within-cluster sum of squared deviation from the cluster means, which is the well-known *K*-means criterion (Hartigan, 1975; MacQueen, 1967).

Alternative adjustments for cluster size can be achieved using different divisors for the within-cluster sums of dissimilarities. One such example is criterion (2.4), which divides each within-cluster sum by the number of pairwise dissimilarity terms used in the computation of that sum. Other possible standardization terms are mentioned by Hubert et al. (2001, Chapter 3). We limit our analyses in this monograph to standardization of the within-cluster sums in accordance with (2.3). However, extensions for other means of standardization are straightforward.

For small values of *n* and *K*, we might be able to produce optimal solutions for (2.1), (2.2), and (2.3) via exhaustive enumeration of all feasible partitions. In general, however, such a strategy is not computationally feasible. Anderberg (1973) and Hand (1981b) provide the following formula for the number of feasible partitions of *n* objects into *K* clusters:

$$\frac{1}{K!}\sum_{k=0}^{K}(-1)^{k}\binom{K}{k}(K-k)^{n}. \tag{2.5}$$

Table 2.1 provides values of (2.5) for various combinations of *n* and *K*. For example, for a 5-cluster partition of a dissimilarity matrix with $n = 30$ objects, the number of feasible partitions is more than seven quintillion, which absolutely precludes complete enumeration.

Klein and Aronson (1991) describe a branch-and-bound procedure for (2.2), which was subsequently enhanced by Brusco (2003). Branch-and-bound algorithms for (2.3) have been described by Koontz et al. (1975), Hand (1981a), Diehr (1985), and Brusco (2005). Hansen and Delattre (1978) proposed an algorithm for (2.1) that incorporates a branch-and-bound method described by Brown (1972).

In the next subsection, we provide a general description of a branch-and-bound paradigm for the partitioning process. The presentation of the general paradigm most closely resembles the algorithm presented by

Klein and Aronson (1991); however, it is comparable to most other branch-and-bound implementations. We adopted this particular presentation of the paradigm because it has conceptual straightforwardness and breadth of applicability.

Table 2.1. Number of feasible partitions for selected values of n and K.

Number of objects (n)	Number of clusters (K)	Number of feasible partitions
10	2	511
	3	9,330
	4	34,105
	5	42,525
	6	22,827
20	2	524,287
	3	580,606,446
	4	45,232,115,901
	5	749,206,090,500
	6	4,306,078,895,384
30	2	536,870,911
	3	34,314,651,811,530
	4	48,004,081,105,038,304
	5	7,713,000,216,608,565,248
	6	299,310,102,746,948,632,576
40	2	549,755,813,887
	3	2,026,277,026,753,674,240
	4	50,369,882,873,307,917,713,408
	5	75,740,854,251,732,104,426,553,344
	6	18,490,198,597,752,082,317,124,304,896

2.2 A Branch-and-Bound Paradigm for Partitioning

Assuming that the n objects have been enumerated into an object list, a branch-and-bound paradigm for partitioning builds solutions by adding successive objects to available clusters. A partial solution consists of an assignment of the first p objects to clusters. The remaining $n - p$ objects have not been assigned a cluster. We denote S_p as the set of objects that have been assigned to clusters and $\overline{S}_p = S \setminus S_p$ (i.e., the complement of S_p with respect to S) as the unassigned objects. The evaluation of a partial solution requires an answer to the question: Is it definitely impossible to assign the objects in $\overline{S}_p$ to clusters so as to provide an optimal solution?

If we can definitively answer this question as "Yes," then the current partial solution need not be pursued any further. Thus, the partial solution, as well as all possible partial and complete clustering solutions that stem from it, can be eliminated from further consideration. If the answer to the question is "No," then we move deeper into the search tree by assigning object $p + 1$ to a cluster.

2.2.1 Algorithm Notation

p = a pointer that corresponds to the current unassigned object under consideration for assignment and marks the depth in the search tree;

τ = the number of empty clusters;

n_k = the number of objects in cluster k, for $1 \le k \le K$;

λ = a vector $[\lambda_1, \lambda_2, ..., \lambda_p]$ containing the current partial solution, where λ_j is the cluster to which object j is currently assigned, for $1 \le j \le p$;

λ^* = the vector $[\lambda_1, \lambda_2, ..., \lambda_n]$ of cluster assignments associated with the incumbent (best found) complete solution;

$f_r(\lambda)$ = the value for criterion r corresponding to a partial solution, λ, where $r = 1, 2, 3, 4$ refers to (2.1), (2.2), (2.3), and (2.4), respectively.

$f_r(\lambda^*) = $ the incumbent (best found) objective value for the selected criterion r corresponding to the incumbent complete solution, λ^*.

2.2.2 Steps of the Algorithm

Step 0. INITIALIZE. Establish λ^* and $f_r(\lambda^*)$ using an efficient heuristic method. Set $p = 0$, $\tau = K$, $\lambda_j = 0$ for $1 \leq j \leq n$; and $n_k = 0$ for $1 \leq k \leq K$.

Step 1. BRANCH FORWARD. Set $p = p + 1$, $k = 1$, $n_k = n_k + 1$, $\lambda_p = k$. If $n_k = 1$, then set $\tau = \tau - 1$.

Step 2. FEASIBILE COMPLETION TEST. If $n - p < \tau$, go to Step 5.

Step 3. PARTIAL SOLUTION EVALUATION. If passed, then go to Step 4. Otherwise, go to Step 5.

Step 4. COMPLETE SOLUTION? If $p \neq n$, then go to Step 1. Otherwise, set $\lambda^* = \lambda$ and store $f_r(\lambda^*)$.

Step 5. DISPENSATION. If $k = K$ or $n_k = 1$, then go to Step 7.

Step 6. BRANCH RIGHT. Set $n_k = n_k - 1$. Set $k = k + 1$ and $n_k = n_k + 1$. If $n_k = 1$, then set $\tau = \tau - 1$. Set $\lambda_p = k$ and return to Step 2.

Step 7. RETRACTION. Set $\lambda_p = 0$. Set $n_k = n_k - 1$ and $p = p - 1$. If $n_k = 0$, then set $\tau = \tau + 1$. If $p = 0$, then return the incumbent solution, which is an optimal solution, and STOP; otherwise, set $k = \lambda_p$ and return to Step 5.

2.2.3 Algorithm Description

The INITIALIZE step of the algorithm (Step 0) can consist of a variety of operations. At a minimum, arrays and variables are set to their appropriate starting values; however, more sophisticated operations might be valuable. Regardless of the criterion, an initial solution that provides a tight upper bound is beneficial because this can greatly reduce the number of partitions that must be explicitly evaluated. Therefore, efficient heuristic procedures are often used to produce initial bounds (Brusco,

2003; Diehr, 1985). Another part of the initialization process might include a reordering of the rows and columns of **A** to enable partial solutions to be eliminated earlier in the branching process. Because the appropriate heuristic procedures and matrix reordering strategies can differ across criteria, we defer these until specific coverage of the criteria.

Step 1 is referred to as "BRANCH FORWARD" because, at this step, the branch-and-bound algorithm is moving deeper into the search tree by assigning a cluster membership to a newly selected object. Step 1 advances the pointer, p, and places the corresponding object in the first cluster ($k = 1$).

Step 2 is a FEASIBILITY TEST that determines whether or not the current partial solution could possibly yield a partition upon completion of the cluster assignments. If $n - p < \tau$, then there are not enough unassigned objects to fill the remaining clusters. Consider, for example, a clustering problem for which $n = 20$ and $K = 6$. If the current pointer position is $p = 17$ and the 17 assigned objects have only been assigned to two of the six clusters, then $\tau = 6 - 2 = 4$. The assignment of the remaining three unassigned objects could only provide memberships for three of the four empty clusters. Thus, it is impossible to complete a six-cluster partition and the current partial solution is pruned.

PARTIAL SOLUTION EVALUATION in Step 3 is the component of the algorithm that really "makes or breaks" the success of the branch-and-bound paradigm. This is the point in the algorithm where the question is posed: Can completion of the partial solution possibly achieve an objective value that is better than the incumbent? The evaluation of a partial solution obviously includes information from the current partial solution itself but, in order to achieve maximum effectiveness, must also reflect what objective criterion contributions can possibly be achieved from the yet unassigned objects. This is perhaps the most challenging aspect of branch-and-bound algorithm design, and appropriate strategies can vary across different criteria.

If the current partial solution passes the tests in Steps 2 and 3, then it is tested for completeness in Step 4. In other words, a check is made to determine whether the partial solution is actually a complete solution. If the partial solution is, in fact, a completed partition, then the partial solution evaluation of Step 3 has determined it to be better than the incumbent solution and, so, it replaces the current incumbent solution ($\lambda^* = \lambda$). Otherwise, the search moves deeper into the tree by returning to Step 1.

A DISPENSATION of the current partial solution is made in Step 5. Step 5 determines whether the current partial solution should BRANCH RIGHT in Step 6 by assigning the current object p to the next cluster ($k +$

1) or whether RETRACTION should occur in Step 7. The conditions in Step 5 are subtle but important because they preclude the evaluation of redundant solutions (Klein & Aronson, 1991). The first condition in Step 5 is intuitive. If the current cluster membership of object p is K, then all clusters have been evaluated for p. Therefore, the algorithm initiates retraction because it cannot assign p to a cluster $K + 1$ that does not exist.

The second condition in Step 5 requires a bit more elaboration. In practice, this condition ensures that the clusters become nonempty in a sequential manner. That is, cluster $k + 1$ is not a candidate and will not be assigned an object unless clusters 1 through k are nonempty. If $n_k = 1$, then moving object p to cluster $k + 1$ will obviously leave cluster k empty. Further, if cluster $k + 1$ was empty prior to moving object p, this means that none of the objects $1, 2,..., p - 1$ are assigned to cluster $k + 1$. Moreover, it means that the objects $1, 2,..., p - 1$ are assigned to clusters $1,..., k - 1$. Although an object $j > p$ could ultimately be assigned to cluster k, that object will be the first object assigned to cluster k, whereas object p will be the first object assigned to cluster $k + 1$. Evaluation of such a partial solution is unnecessary because it was already considered when p was in cluster k. Simply put, the second condition in Step 5 is an implicit way of avoiding evaluation of all $K!$ configurations of cluster assignments that differ only in the labeling of the clusters, not the cluster memberships themselves.

The BRANCH RIGHT step (Step 6) moves the object from its current cluster k and updates n_k accordingly. The cluster index is incremented to $k + 1$ and the object is placed into this "next" cluster. If this is the first object in the cluster, then τ is decremented accordingly ($\tau = \tau - 1$). Processing subsequently returns to Step 2 for testing of this new partial solution. Consider, for example, a $K = 4$ cluster problem and a current pointer position of $p = 8$, where $\lambda_8 = 3$, and $n_k = 2$ for $k = 1, 2, 3, 4$. At Step 6, the following sequence of changes are applied: $n_3 = n_3 - 1 = 1$, $\lambda_8 = \lambda_8 + 1 = 4$, and $n_4 = n_4 + 1 = 3$. Because clusters 3 and 4 were both nonempty prior to and after the branch right step, there is no change in the value of τ.

RETRACTION in Step 7 begins with the removal of the object corresponding to the pointer p from its current cluster and decrementing n_k. If this results in cluster k becoming empty, then the value of τ is incremented. Retraction occurs by decrementing p (i.e., backing up in the object list). If $p > 0$ then the algorithm proceeds by setting $k = \lambda_p$ and proceeding to Step 5 for dispensation. If $p = 0$, enumeration is complete and the algorithm terminates with the return of the optimal solution.

The main algorithm described in this chapter is applicable to the cluster analysis problems that are of interest to us in this monograph. The components that are specific to particular problems are the heuristics in the INITIALIZE step and the PARTIAL SOLUTION EVALUATION. The algorithmic pseudocode for the main algorithm can be found in Appendix A and the individual routines are described in the discussions of particular partitioning problems.

Throughout the remainder of the chapters discussing cluster analysis, some standard terminology will be used in the pseudocode presentations. The term $\mathbf{A}(i, j)$ will correspond directly to elements of the dissimilarity matrix $\mathbf{A} = \{a_{ij}\}$. The number of clusters, K, will be denoted by *num_k* and a particular cluster will typically be referred to using k or other lowercase letters such as h. Partitioning solutions, λ, will be addressed by the vector lambda(i). The term incumbent will refer to the incumbent criterion (or diameter) value $f_r(\lambda^*)$.

3 Minimum-Diameter Partitioning

3.1 Overview

Partitioning based on the diameter criterion (2.1) has a rich history in the classification literature. The diameter criterion is also known as the Maximum Method (Johnson, 1967) due to the focus on the minimization of the maximum dissimilarity element within clusters. Compact Partitioning is also a viable descriptor for minimum-diameter partitioning because the resultant clusters are kept as tight as possible given the downward pressure on the maximum dissimilarities within clusters.

There are fundamental relationships between minimum-diameter partitioning, complete-link hierarchical cluster analysis, and the coloring of threshold graphs, and these are well documented in the classification literature (Baker & Hubert, 1976; Brusco & Cradit, 2004; Guénoche, 1993; Hansen & Delattre, 1978; Hubert, 1974). Complete-link hierarchical clustering is an agglomerative method that, at each level of the hierarchy, merges clusters that result in the minimum increase in the maximum pairwise dissimilarity element. Cutting the tree at some number of clusters $K \geq 2$ can produce a partition of objects; however, the resulting partition is not guaranteed to be an optimal solution for (2.1). In fact, results from Hansen and Delattre (1978) suggest that the suboptimality associated with the use of complete-link hierarchical clustering as a heuristic for the minimum-diameter partitioning problem can be quite severe.

Rao (1971) demonstrated that, for the special case of $K = 2$, the minimum-diameter partitioning problem (2.1) can be solved using a straightforward repulsion algorithm. This bipartitioning procedure has also been discussed by Hubert (1973), and a streamlined algorithm was presented by Guénoche et al. (1991). The divisive bipartitioning approach can also be used, in a heuristic manner, to produce solutions to (2.1) for $K \geq 3$. Specifically, using K-1 bipartition splits (each time focusing on the cluster with the largest diameter), a K-cluster partition is produced. The final partition obtained, however, does not necessarily yield a minimum diameter when $K \geq 3$.

Hansen and Delattre (1978) proved that (2.1) is NP-hard for $K \geq 3$. They also expounded on Hubert's (1974) earlier observation of the linkage between the diameter criterion and the coloring of a threshold graph, and presented an algorithm for (2.1) that incorporates a branch-and-bound procedure developed by Brown (1972) within the context of chromatic scheduling. Like most branch-and-bound algorithms for clustering, the performance of the algorithm is sensitive to the properties of the dissimilarity matrix. However, the algorithm is often able to produce optimal solutions for problems with hundreds of objects, particularly for smaller values of K.

3.2 The INITIALIZE Step

The algorithm described in Chapter 2 performs extremely well for (2.1), provided that special care is taken in the implementation of the algorithm. For example, the INITIALIZE step (Step 0) advisably uses an efficient heuristic procedure to produce a tight upper bound. A natural choice is to apply complete-link hierarchical clustering and cut the tree at the desired number of clusters (see Baker and Hubert [1976] for a discussion of complete-link hierarchical clustering). However, as noted previously, Hansen and Delattre (1978) observed that a complete-link solution often yields a partition diameter that exceeds the optimum and, in some instances, the departure from optimality is severe.

Better upper bounds can be obtained by using replications of the complete-link algorithm with biased sampling in the selection of linkages. With this tactic, suboptimal linkages can be accepted (with small probability) during the construction of the tree. The idea is that, in some instances, a suboptimal linkage at a higher level of the tree can ultimately lead to a better diameter for the desired K. Brusco (2003) observed that this type of heuristic provided excellent (often optimal) initial starting solutions in an extremely efficient manner. Another option is to use an exchange algorithm to attempt to improve a complete-link clustering solution. For our implementation of minimum-diameter partitioning, we use a complete linkage algorithm followed by an exchange procedure that attempts to move objects across clusters and accepts any move that improves partition diameter.

The complete-link hierarchical clustering algorithm begins with each object in its own individual cluster (i.e., there are n clusters, each containing one object). Each iteration of the algorithm merges a pair of clusters, with the objective of keeping the increase in the diameter resulting from the merger as small as possible. Thus, after one, two, and three iterations, there are $n-1$, $n-2$, and $n-3$ clusters, respectively. The complete-link procedure continues until there are exactly K clusters. The algorithm is sensitive to the order of the data, as well as ties in the data set. For most problems, a minimum-diameter partition solution for K clusters is not realized from the algorithm. Therefore, we recommend applying the exchange algorithm depicted in the pseudocode below, which is similar to the strategy described by Banfield and Bassil (1977), immediately following the complete-link algorithm. The exchange algorithm consists of two phases: (a) single-object relocation, and (b) pairwise interchange. The single-object relocation phase examines the effect of moving each object from its current cluster to each of the other clusters. Any move that improves the diameter is accepted. The pairwise interchange phase evaluates all possible swaps of objects that are not in the same cluster. The two phases are implemented until no relocation or interchange further improves the diameter criterion. The resulting solution, although not necessarily a global optimum, is locally optimal with respect to relocations and interchanges. We now consider pseudocode for an initialization procedure.

{**Randomly Generate** a feasible partition of the n objects into *num_k* clusters and let lambda(i) represent the cluster membership of object i, for $1 \le i \le n$.}
Set flag = True
while flag
  set flag = False; flag1 = True
  while flag1   /* SINGLE OBJECT RELOCATION */
    flag1 = False
    for i = 1 to *n*
      h = lambda(i)
      diam_h = 0
      for j = 1 to *n*
        if lambda(j) = h and i <> j then
          if **A**(i, j) > diam_h then diam_h = **A**(i, j)
        end if
      next j
      for k = 1 to *num_k*
        diam_k = 0
        if k <> h then
          for j = 1 to n
            if lambda(j) = k and i <> j then

```
            if A(i, j) > diam_k then diam_k = A(i, j)
            end if
          next j
          if diam_h > diam_k then
            lambda(i) = k, h = k, diam_h = diam_k
            flag1 = True, flag = True
          end if
        end if
      next k
    next i
  loop
  flag2 = True
  while flag2                        /* PAIRWISE INTERCHANGE */
    flag2 = False
    for i = 1 to n − 1
      for m = i + 1 to n
        h = lambda(i), q = lambda(m)
        if h <> q then
          diam_ih = 0: diam_iq = 0: diam_jh = 0: diam_jq = 0
          for Position = 1 to n
            if lambda(Position) = h then
              if Position <> i and A(i, Position) > diam_ih then
                diam_ih = A(i, Position)
              if Position <> j and A(j, Position) > diam_jh then
                diam_jh = A(j, Position)
            end if
            if lambda(Position) = q then
              if Position <> i and A(i, Position) > diam_iq then
                diam_iq = A(i, Position)
              if Position <> j and A(j, Position) > diam_jq then
                diam_jq = A(j, Position)
            end if
          next Position
          if max(diam_jh, diam_iq) − max(diam_ih, diam_jq) < 0 then
            lambda(i) = q, lambda(j) = h and flag2 = True, flag = True
          end if
        end if
      next m
    next i
  loop                  /* End of pairwise interchange loop */
loop                    /* End of exchange algorithm */
```

3.2 The INITIALIZE Step

Another important component of the INITIALIZE process for (2.1) is the reordering of objects prior to implementation of the algorithm. The rationale is that, by placing objects with a large number of matrix elements that equal or exceed the diameter estimate early in the list, the incumbent solution will be sufficiently small to eliminate a large number of partial solutions early in the branch-and-bound tree. We have adopted a procedure that reorders the objects based on the number of pairwise elements that equal or exceed the partition diameter obtained by the heuristic solution procedure, $f_1(\lambda^*)$. Specifically, for each object i, the following value is computed:

$$\alpha_i = \sum_{j \neq i} \omega_{ij}, \qquad (3.1)$$

$$\text{where: } \omega_{ij} = \begin{cases} 1 & \text{if } a_{ij} \geq f_1(\lambda^*) \\ 0 & \text{otherwise.} \end{cases}$$

Objects are reordered based on descending values of α_i. The algorithmic pseudocode for this portion of the INITIALIZE step uses the initial incumbent found by the previous heuristic.

```
for i = 1 to n
   alpha(i) = 0                          /* Compute alpha(i) */
   for j = 1 to n
      if i <> j then
         if A(i, j) >= incumbent_diameter then alpha(i) = alpha(i) + 1
      end if
   next j
next i
for i = 1 to n
   order(i) = i
next i
for i = 1 to n – 1                       /* Sort based on largest alpha(i) */
   for j = i + 1 to n
      if alpha(j) > alpha(i) then
         dummy = alpha(i), alpha(i) = alpha(j), alpha(j) = dummy
         dummy = order(i), order(i) = order(j), order(j) = dummy
      end if
   next j
next i
```

```
for i = 1 to n          /* Reorder Matrix A */
  for j = 1 to n
    M(i, j) = A(order(i), order(j))
  next j
next i
for i = 1 to n
  for j = 1 to n
    A(i, j) = M(i, j)
  next j
next i                  /* End of reordering subroutine */
```

3.3 The PARTIAL SOLUTION EVALUATION Step

There are two components to the evaluation of partial solutions in Step 3 of the branch-and-bound process for (2.1). The first test in the evaluation of the partial solution corresponds to objects that have already been assigned a cluster membership. Whenever the object in position p is assigned to a new cluster k, we evaluate whether the inclusion of the object in the cluster will create a within-cluster pairwise dissimilarity that equals or exceeds the incumbent partition diameter value $f_1(\lambda^*)$. More formally, a partial solution created by assigning object p to cluster k is pruned if:

$$[a_{ip} \geq f_1(\lambda^*) \mid \lambda_i = k], \text{ for any } i = 1,..., p-1. \tag{3.2}$$

The second component of PARTIAL SOLUTION EVALUATION for (2.1) corresponds to objects $p + 1,..., n$, which have yet to be assigned a cluster membership. This test, which is only performed if all clusters have been assigned at least one object (that is, if $\tau = 0$), is based on the fact that it might not be possible to assign one of the yet unassigned objects to *any* cluster without creating a within-cluster pairwise dissimilarity that equals or exceeds the incumbent partition diameter. More formally, a partial solution created by assigning object p to cluster k is pruned if:

$$\min_k \left[\max_{q=1,...,p} \left(a_{jq} \mid \lambda_q = k \right) \right] \geq f_1(\lambda^*), \text{ for any } j = p+1,..., n. \tag{3.3}$$

The unassigned objects are evaluated in turn and, if any object cannot be assigned to at least one of the K clusters without creating a pairwise dissimilarity value that equals or exceeds the incumbent partition diame-

3.3 The PARTIAL SOLUTION EVALUATION Step

ter, then the solution is pruned. If we are attempting to assign object p (*Position*) to cluster k, then the algorithmic pseudocode for the PARTIAL SOLUTION EVALUATION incorporates both components (3.2) and (3.3) and uses the diameter of the incumbent solution as the upper bound.

Set Prune = False, Fathom = True and PARTIAL_EVAL = Fathom.
/* **Determine** whether or not object in *Position* will give cluster k an unacceptably high diameter. */
for i = 1 to *Position* – 1
  if lambda(i) = k and **A**(i, *Position*) >= *incumbent* then
    PARTIAL_EVAL = Prune
  end if
next i
/* If all clusters currently have at least one object and the partial solution has not been previously deemed unacceptable, then **determine** whether or not assigning any of the remaining objects to a cluster will necessarily result in an unacceptably high diameter. */
if PART_EVAL and tau = 0 then
  Unassigned = *Position* + 1
  while PART_EVAL and Unassigned <= n
    /* For each cluster, **find** the increase to the diameter that would
       result from adding the unassigned object */
    for CheckCluster = 1 to *num_k*
      MaxD = 0
      for Assigned = 1 to *Position*
        if cluster(Assigned) = CheckCluster then
          if MaxD < **A**(Unassigned, Assigned) then
            MaxD = **A**(Unassigned, Assigned)
        end if
      next Assigned
      diameter(CheckCluster) = MaxD
    next CheckCluster
  /* **Find** the cluster that would result in the minimum diameter
    increase. */
    MinD = diameter(1)
    for MinK = 2 to *num_k*
      if MinD > diameter(MinK) then MinD = diameter(MinK)
    next MinK

       /* **Determine** whether the minimum diameter increase would be
         acceptable.*/
       if MinD > *incumbent* then PART_EVAL = Prune
       Unassigned = Unassigned + 1
    loop   /* End of Unassigned loop */
  end if   /* End of partial evaluation test */

3.4 A Numerical Example

To illustrate the application of the branch-and-bound algorithm to minimum-diameter partitioning, we consider the 6 × 6 dissimilarity matrix shown in Table 3.1. We will assume that the INITIALIZE step has employed a heuristic that yields the $K = 2$ cluster partition $\{1, 2, 5\}$, $\{3, 4, 6\}$ or equivalently $\lambda^* = [1, 1, 2, 2, 1, 2]$, which provides an initial upper bound for partition diameter of $f_1(\lambda^*) = 53$.

Table 3.1. Example dissimilarity matrix #1.

	1	2	3	4	5	6
1	----					
2	35	----				
3	59	17	----			
4	52	88	43	----		
5	24	50	63	37	----	
6	47	41	53	29	34	----

The iterations of the branch-and-bound algorithm for a 2-cluster partition are shown in Table 3.2. The first two columns provide the row label and the cluster assignments for each object (e.g., the cluster assignment in row of 1 1 2 in row 4 indicates that objects 1 and 2 are assigned to cluster 1, object 3 is assigned to cluster 2, and objects 4, 5, and 6 are not yet assigned). These are followed by columns for the pointer, p, the diameters of clusters $k = 1$ and $k = 2$, and the dispensation of the partial solution. The initial incumbent solution is $f_1(\lambda^*) = 53$. Notice that the final invocation of DISPENSATION leads to the termination of the algorithm because assigning the first object to the second cluster could only result in a redundant solution, i.e. all objects from the previous solution assigned to cluster 1 would be assigned to cluster 2 and vice versa. This pruning takes place in Step 5 of the main algorithm because $k = K$ and the termination follows in the RETRACTION of Step 7.

Table 3.2. Branch-and-bound summary for the data in Table 3.1.

			Diameters		
Row	λ	p	$k = 1$	$k = 2$	Dispensation
1	1	1	0	0	Branch forward
2	1 1	2	35	0	Branch forward
3	1 1 1	3	59	0	Prune $(59 \geq 53)$, Branch right
4	1 1 2	3	35	0	Branch forward
5	1 1 2 1	4	88	0	Prune $(88 \geq 53)$, Branch right
6	1 1 2 2	4	35	43	Branch forward
7	1 1 2 2 1	5	50	43	Branch forward
8	1 1 2 2 1 1	6	50	43	*New Incumbent, $f_1(\lambda^*) = 50$
9	1 1 2 2 1 2	6	50	53	Prune $(53 \geq 50)$, Retract
10	1 1 2 2 2	5	35	63	Prune $(63 \geq 50)$, Retract
11	1 2	2	0	0	Prune (Component 2), Retract
12	2	1	0	0	TERMINATE

The first pruning of the branch-and-bound tree occurs in row 3, where the partial solution corresponding to the assignment of objects 1, 2, and 3 to the first cluster is eliminated because $a_{13} = 59 \geq f_1(\lambda^*) = 53$. A new incumbent solution with diameter $f_1(\lambda^*) = 50$ is identified in row 8, and corresponds to the partition $\{1, 2, 5, 6\}, \{3, 4\}$. A particularly efficient pruning operation occurs in row 11 of the table for the partial solution associated with the assignments of object 1 to cluster 1 and object 2 to cluster 2. This partial solution is pruned with the deployment of the second component of PARTIAL SOLUTION EVALUATION in (3.3). If object 4 is placed in cluster 1, then $a_{14} = 52$, whereas if object 4 is placed in cluster 2, then $a_{24} = 88$. Because object 4 must be placed in one of these two clusters, a within-cluster pairwise dissimilarity of at least 52 would be realized, which exceeds the incumbent partition diameter of $f_1(\lambda^*) = 50$.

3.5 Application to a Larger Data Set

Although the example in the previous subsection is sufficiently small to facilitate the demonstration of the branch-and-bound algorithm, the algorithm is scalable for much larger matrices of greater pragmatic interest. We consider an example based on similarity data pertaining to recognition of lipread consonants. These data were originally collected and analyzed by Manning and Shofner (1991) and have also been used by Brusco and Stahl (2004) and Brusco and Cradit (2005) for purposes of demonstrating combinatorial data analysis methods. We converted the similarity matrix reported by Manning and Shofner (1991, p. 595) to the symmetric dissimilarity matrix shown in Table 3.3 by taking 2 minus the similarity measures and multiplying by 100, which resulted in a non-negative 21 × 21 matrix with elements ranging from 49 to 365.

Table 3.3. Dissimilarity matrix for lipread consonants based on data from Manning and Shofner (1991).

	b	c	d	f	g	h	j	k	l	m	n
b		141	176	308	155	118	265	296	298	331	280
c	141		118	292	149	280	306	229	194	325	265
d	176	118		251	147	288	235	273	227	324	196
f	308	292	251		298	271	282	275	249	216	243
g	155	149	147	298		273	157	269	290	324	241
h	118	280	288	271	273		267	269	233	267	200
j	265	306	235	282	157	267		182	241	322	269
k	296	229	273	275	269	269	182		184	296	204
l	298	194	227	249	290	233	241	184		175	149
m	331	325	324	216	324	267	322	296	175		243
n	280	265	196	243	241	200	269	204	149	243	
p	69	214	149	290	275	282	275	288	269	296	243
q	284	312	327	331	271	306	300	247	316	312	255
r	318	314	321	296	300	280	324	214	202	302	243
s	149	292	245	204	284	249	286	271	147	276	176
t	182	122	90	288	169	288	298	229	271	304	210
v	227	182	173	265	173	292	275	292	345	296	269
w	355	325	335	355	312	345	349	318	361	316	347
x	282	255	286	255	298	178	286	198	176	271	214
y	308	329	337	316	331	325	286	325	320	335	308
z	129	49	82	311	182	249	278	275	292	292	273

Table 3.3. – Continued

	p	q	r	s	t	v	w	x	y	z
b	69	284	318	149	182	227	355	282	308	129
c	214	312	314	292	122	182	325	255	329	49
d	149	327	321	245	90	173	335	286	337	82
f	290	331	296	204	288	265	355	255	316	311
g	275	271	300	284	169	173	312	298	331	182
h	282	306	280	249	288	292	345	178	325	249
j	275	300	324	286	298	275	349	286	286	278
k	288	247	214	271	229	292	318	198	325	275
l	269	316	202	147	271	345	361	176	320	292
m	296	312	302	276	304	296	316	271	335	292
n	243	255	243	176	210	269	347	214	308	273
p		308	302	286	176	200	365	312	337	180
q	308		192	284	265	294	298	327	284	251
r	302	192		300	306	302	337	227	363	271
s	286	284	300		269	292	363	75	331	208
t	176	265	306	269		131	349	220	347	147
v	200	294	302	292	131		329	251	310	204
w	365	298	337	363	349	329		334	300	345
x	312	327	227	75	220	251	334		298	289
y	337	284	363	331	347	310	300	298		296
z	180	251	271	208	147	204	345	239	296	

We used the branch-and-bound procedure to produce optimal minimum-diameter partitions for the data in Table 3.3 for values of $2 \leq K \leq 8$. The optimal partition diameters for the various numbers of clusters are displayed in Table 3.4. There is no definitive rule for selecting the appropriate number of clusters based on partition diameter; however, a *diameter reduction ratio* has been recommended by Hansen and Delattre (1978). In particular, this criterion is the percentage reduction in diameter obtained from increasing the number of clusters by one. We denote the diameter reduction ratio obtained by increasing the number of clusters from $K-1$ to K as:

$$drr(K) = 100\left(\frac{f_1^*(\pi_{K-1}) - f_1^*(\pi_K)}{f_1^*(\pi_{K-1})}\right). \tag{3.4}$$

3 Minimum-Diameter Partitioning

Table 3.4. Results for lipread consonants data at different numbers of clusters (K).

K	Partition diameter	$drr(K)$	Partition
2	331	--	{b,c,d,f,g,h,j,k,*l*,m,n,p,r,s,t,x,z} {q,v,w,y}
3	306	7.55	{b,c,d,g,j,p,t,v,z} {f,h,k,*l*,m,n,r,s,x} {q,w,y}
4	302	1.31	{b,c,d,p,t,v,z} {f,k,*l*,m,n,r} {g,h,j,s,x} {q,w,y}
5	286	5.29	{b,c,d,g,p,t,v} {f,h,*l*,m,n,s,x} {k,q,r,z} {j,y} {w}
6	276	3.50	{b,c,d,p,t,v,z} {f,h,*l*,m,n,s,x} {g,j} {k,q,r} {w} {y}
7	249	9.78	{b,c,d,p,t,v,z} {f,m} {g,j} {h,*l*,n,s,x} {k,q,r} {w} {y}
8	247	0.80	{b,c,d,p,t,v,z} {f,m} {g,j} {h} {k,q,r} {*l*,n,s,x} {w} {y}

The $drr(K)$ values in Table 3.4 indicate that $K = 3$ or $K = 7$ clusters might be good choices for the number of clusters. There is a 7.55% reduction in diameter when increasing K from 2 to 3, but only a 1.31% reduction from increasing K from 3 to 4. Similarly, there is a 9.78% reduction in diameter when increasing K from 6 to 7, but only a .80% reduction from increasing K from 7 to 8.

The within-cluster submatrices for the optimal minimum-diameter partitioning solution for $K = 3$ clusters are presented in Table 3.5. Two of the clusters have nine objects and the third cluster has only three objects; however, the cluster diameters are similar across the three clusters. The diameter for cluster 1 is 306 (contributed by object pair (c, j)), which defines the partition diameter because it is the largest of the cluster diameters. The diameter for cluster 2 is 302 (contributed by object pair (m, r)), and the diameter for cluster 3 is 300 (contributed by object pair (w, y)). Cluster 1 {b, c, d, g, j, p, t, v, z} contains mostly letters with a "long e" sound (letter "j" is the exception). Even though the letters are only lipread (not heard), we can infer that these letters might be somewhat confusable. Further, the confusion of "j" with "g" when lipread seems intuitive. Cluster 2 {f, h, k, l, m, n, r, s, x} contains mostly letters that have an "eh" sound at the beginning (e.g., "ef", "el", "em", "en", "es", "ex"), and cluster 3 {q, w, y} consists of letters that require some degree of puckering the lips when mouthed. In summary, the three-cluster solution has at least some degree of practical interpretability. However, as we will later see, this is not the only 3-cluster minimum-diameter partition.

Table 3.5. Within-cluster submatrices for lipread consonants data in Table 3.3 that are obtained when optimizing (2.1).

Cluster 1

	b	c	d	g	j	p	t	v
b	--							
c	141	--						
d	176	118	--					
g	155	149	147	--				
j	265	306	235	157				
p	69	214	149	275	275	--		
t	182	122	90	169	298	176	--	
v	227	182	173	173	275	200	131	--
z	129	49	82	182	278	180	147	204

Cluster 2

	f	h	k	l	m	n	r	s
f	--							
h	271	--						
k	275	269	--					
l	249	233	184	--				
m	216	267	296	175	--			
n	243	200	204	149	243	--		
r	296	280	214	202	302	243	--	
s	204	249	271	147	276	176	300	--
x	255	178	198	176	271	214	227	75

Cluster 3

	q	w	y
q	--		
w	298	--	
y	284	300	--

3.6 An Alternative Diameter Criterion

We have focused our discussion on the minimization of the maximum of the cluster diameters (i.e., the partition diameter). There are many other possibilities for establishing criteria based on cluster diameters, perhaps the most obvious of which is the minimization of the *sum of the cluster diameters*:

$$\min_{\pi_K \in \Pi_K} : f_5(\pi_K) = \sum_{k=1}^{K} \left(\max_{(i<j) \in C_k} (a_{ij}) \right). \qquad (3.5)$$

Brucker (1978) showed that the problem posed by (3.5) is NP-hard for $K \geq 3$. Polynomial algorithms for optimal bipartitioning solutions ($K = 2$) for (3.5) have been developed by Hansen and Jaumard (1987) and Ramnath, Khan, and Shams (2004).

We adapted the branch-and-bound algorithm for minimizing the sum of diameters. The necessary modifications for partial solution evaluation are relatively minor. However, the bounding procedure is not as sharp and the algorithm for minimizing the sum of cluster diameters does not have the scalability of the algorithm for minimizing partition diameter. We applied the algorithm for minimizing the sum of cluster diameters to the data in Table 3.1 assuming $K = 2$ and obtained the partition {1, 2, 3, 5, 6}, {4}. The diameter for the cluster {1, 2, 3, 5, 6} is 63 and the diameter for {4} is zero, resulting in a sum of diameters index of 63 and a partition diameter of 63. Recall that the optimal partition diameter of 50 is obtained from the partition {1, 2, 5, 6} {3, 4}, which has a sum of cluster diameters of 50 + 43 = 93. Thus, the optimal solutions for the two diameter criteria are not the same. We also make the worthwhile observation that the partition {1, 4, 5, 6} {2, 3}, which in not an optimal solution for either criterion, is nevertheless an excellent compromise solution. The diameter for the cluster {1, 4, 5, 6} is 52 and the diameter for {2, 3} is 17, resulting in a sum of diameters index of 69 and a partition diameter of 52.

We have observed that the minimization of the sum of cluster diameters has a tendency to produce a fairly large number of "singleton" clusters (that is, clusters with only one object). Clusters with only one object produce cluster diameters of zero, and this property often promotes the peeling off of objects into singleton clusters. Although this tendency might limit the utility of the sum of diameters criterion for some applications, the criterion nevertheless can facilitate interesting comparisons with the more popular partition diameter criterion.

3.7 Strengths and Limitations

One of the advantages of the partition diameter index is that, unlike (2.2), (2.3), and (2.4), it is not predisposed to produce clusters of particular sizes. This is important for contexts with the potential for one fairly large cluster of objects and a few small clusters. Another advantage is that minimization of the partition diameter is computationally less difficult than minimizing the within-cluster sums of dissimilarities. The pruning rules are very strong and, particularly when using a good reordering of the objects, optimal solutions can often be obtained for problems with hundreds of objects and ten or fewer clusters.

One of the limitations of the minimum-diameter criterion is that it tends to produce a large number of alternative optimal solutions. The alternative optima can often differ markedly with respect to membership assignments, as well as relative cluster sizes. The quantitative analyst must, therefore, select an appropriate optimal solution from the candidate pool. There are at least two alternatives for facilitating this task. One approach is to apply a method that enumerates the complete set (or a subset of the complete set) of minimum-diameter partitions (Guénoche, 1993). Another approach is to apply multicriterion clustering procedures to break ties among the minimum-diameter partitions based on a secondary criterion (Brusco & Cradit, 2005; Delattre & Hansen, 1980). We will discuss this latter approach in Chapter 6.

3.8 Available Software

We have made available two software programs for diameter-based partitioning, which can be found at http://garnet.acns.fsu.edu/~mbrusco or http://www.psiheart.net/quantpsych/monograph.html. The first program, *bbdiam.for*, is designed for minimizing (2.1). This program closely resembles the algorithm described by Brusco and Cradit (2004). A heuristic is used to establish an initial bound for partition diameter, and is also used to favorably reorder the objects prior to implementation of the branch-and-bound process. The program reads a data file "*amat.dat*" which contains a matrix of pairwise dissimilarities or distances. The first line of the data file should contain the number of objects n, and succeeding lines should contain the matrix data. The user will be prompted for the form of the matrix, and should enter 1 for a half matrix (lower triangle) or 2 for a full $n \times n$ matrix. The user also will be prompted for the desired number of clusters, which can range from 2 to 20. The output of the program is written both to the screen and a file *"results."* The output includes the partition diameter corresponding to the heuristic solution (the upper bound), optimal partition diameter, the CPU time required to obtain the solution, and the cluster assignments for each object. The program *"bbdisum.for"* has the same file structures as *bbdiam.for*, but is designed for minimization of the sum of the cluster diameters.

To illustrate the operation of *bbdiam.for*, we present the following screen display information for execution of the program for the data in Table 3.3 at $K = 3$ clusters.

```
> TYPE 1 FOR HALF MATRIX OR TYPE 2 FOR FULL MATRIX
> 1
> PLEASE INPUT NUMBER OF CLUSTERS 2 TO 20
> 3
> HEURISTIC SOLUTION DIAMETER        306.00000
> THE OPTIMAL MINIMUM DIAMETER       306.00000
> THE TOTAL COMPUTATION TIME              0.01
  1  1  1  2  1  2  1  2  2  2  2  1  3  2  2
  1  1  3  2  3  1
Stop - Program terminated.
```

For this particular data set, the heuristic produces the optimal partition diameter and thus the initial upper bound is very tight. The optimal cluster assignments are written in row form. Objects 1, 2, 3, 5, 7, 12, 16, 17, and 21 are assigned to cluster 1. For these data, those objects correspond to {b, c, d, g, j, p, t, v, z}, as described in section 3.5 above.

The application of *bbdisum.for* to the data in Table 3.3 under the assumption of $K = 3$ clusters yields the following results:

```
> TYPE 1 FOR HALF MATRIX OR TYPE 2 FOR FULL MATRIX
> 1
> PLEASE INPUT NUMBER OF CLUSTERS 2 TO 20
> 3
> THE OPTIMAL MINIMUM SUM OF DIAMETERS      345.00000
       1          0.0000
       2          0.0000
       3        345.0000
  THE TOTAL COMPUTATION TIME                   0.03
     3  3  3  3  3  3  3  3  3  3  3  3  3  3  3  3
     3  1  3  2  3
Stop - Program terminated.
```

The three-cluster partition for the sum of diameters criterion is appreciably different from the three-cluster minimum partition diameter solution. Two of the three clusters consist of only one object, {w} and {y}, respectively. These objects have some rather large dissimilarities with other objects, which is the reason they are placed in their own singleton clusters. Notice that the sum of diameters of 345 is also the partition diameter. Although this partition diameter is somewhat higher than the optimal partition diameter of 306, the minimum partition diameter solution produces a sum of diameters index of 306 + 302 + 300 = 908, which is appreciably larger than 345.

The diameter-based partitioning algorithms are currently dimensioned for up to 250 objects and 20 clusters. Actual problem sizes that can be handled by the algorithm are conditional not only on n and K, but also on the structure of the data. For instances if $K = 2$, we would expect these branch-and-bound implementations to be significantly outperformed by the appropriate bipartitioning algorithms (Guénoche et al., 1991; Hansen & Jaumard, 1987).

4 Minimum Within-Cluster Sums of Dissimilarities Partitioning

4.1 Overview

Klein and Aronson (1991) describe the appropriateness of the within-cluster sums of pairwise dissimilarities criterion (2.2) within the context of computer-assisted process organization, where totality of interaction within clusters is of particular importance. They present an integer linear programming formulation of the problem, but recommended a branch-and-bound procedure as a more plausible solution approach. Brusco (2003) subsequently offered some enhanced bounding procedures that improved the performance of this algorithm. We should note, however, that rather large instances of (2.2) have been successfully solved by methods incorporating cutting planes and column generation procedures (Johnson, Mehrotra, & Nemhauser, 1993; Palubeckis, 1997). The simple and straightforward pseudocode to evaluate any feasible cluster assignment configuration illustrates the criterion to be minimized:

```
/* Compute the within-cluster sum of dissimilarities */
for k = 1 to num_k   /* Initialize sum for each cluster */
   WithinSum(k) = 0
next k
for Pos1 = 1 to n – 1
   k1 = lambda(Pos1)
   for Pos2 = Pos1 + 1 to n
      k2 = lambda(Pos2)
      if k1 = k2 then WithinSum(k1) = WithinSum(k1) + A(Pos1, Pos2)
   next Pos2
next Pos1
EVALUATION = 0
for k = 1 to num_k
   EVALUATION = EVALUATION + WithinSum(k)
next k                      /* End computation for criterion (2.2) */
```

4.2 The INITIALIZE Step

An efficient heuristic is recommended for providing a good upper bound for (2.2). Klein and Aronson (1991) used subroutines from Zupan (1982) to provide an initial incumbent solution; however, the objective functions provided from these subroutines often significantly exceeded the optimal value of (2.2). An exchange algorithm, such as the one described by Banfield and Bassil (1977), is a much preferred alternative. Beginning with an initial feasible solution (often randomly generated), the exchange algorithm modifies the partition until it is locally optimal with respect to two neighborhood operations: (a) no improvement in the objective function can be realized from moving an object from its current cluster to one of the other clusters, and (b) no interchange of objects in different clusters can improve the objective function.

The psuedocode presented in section 3.2, with minor modification for the computation of the change in within-cluster dissimilarity sums instead of diameters, illustrates the basic implementation of the Banfiled and Bassil (1977) procedure.

```
k = 0
for i = 1 to n
    k = k + 1
    if k > num_k then k = 1
    lambda(i) = k
next i
for k = 1 to num_k     /*Find sum of dissimilarities for each cluster */
    WithinSum(k) = 0
next k
for cluster = 1 to num_k
    for object = 1 to n
        if lambda(object) = cluster then
            for object2 = object + 1 to n
                if lambda(object2) = cluster then
                    WithinSum(cluster) = WithinSum(cluster) + A(object, object2)
            next object2
        end if
    next object
next cluster
flag = True
while flag
    Set flag = False and flag1 = True
```

4.2 The INITIALIZE Step

```
while flag1                    /* SINGLE OBJECT RELOCATION */
  flag1 = False
  for object = 1 to n
    for NewCluster = 1 to num_k
      if NewCluster <> lambda(object) then
        NewWithin = WithinSum(lambda(object))
        for bedfellow = 1 to n
          if lambda(object) = lambda(bedfellow) then
            NewWithin = NewWithin - A(object, bedfellow)
        next bedfellow
        NewWithin2 = WithinSum(NewCluster)
        for object2 = 1 to n
          if lambda(object2) = NewCluster then
             NewWithin2 = NewWithin2 + A(object, object2)
        next object2
        if (WithinSum(lambda(object)) + WithinSum(NewCluster))
                            > (NewWithin + NewWithin2) then
          /* Reassign */
          WithinSum(lambda(object)) = NewWithin
          WithinSum(NewCluster) = NewWithin2
          lambda(object) = NewCluster
          flag1 = True
        end if
      end if
    next NewCluster
  next object
Loop     /* End of object relocation, flag1 */
flag2 = True
while flag2                    /* PAIRWISE INTERCHANGE */
  flag2 = False
  for i = 1 to n - 1
    for j = i + 1 to n
      if lambda(i) <> lambda(j) then
        WithinI = WithinSum(lambda(i))
        WithinJ = WithinSum(lambda(j))
        for bedfellow = 1 to n
          if lambda(bedfellow) = lambda(i) then
             WithinI = WithinI - A(i, bedfellow) + A(j, bedfellow)
          if lambda(bedfellow) = lambda(j) then
             WithinJ = WithinJ - A(j, bedfellow) + A(i, bedfellow)
        next bedfellow
```

```
                if (WithinSum(lambda(i)) + WithinSum(lambda(j)))
                                      > (WithinI + WithinJ) then
                    WithinSum(lambda(i)) = WithinI
                    WithinSum(lambda(j)) = WithinJ
                    swap = lambda(i), lambda(i) = lambda(j), lambda(j) = swap
                    flag2 = True
                end if
            end if
        next j
    next i
loop                    /* End of pairwise interchange loop; flag2 */
loop                    /* End of exchange algorithm; flag */
```

The initial upper bound for (2.2) produced by this heuristic process is denoted $f_2(\lambda^*)$. For the remainder of this chapter, this notation refers to the best found solution at any point in the branch-and-bound process.

4.3 The PARTIAL SOLUTION EVALUATION Step

Klein and Aronson's (1991) evaluation of partial solutions requires consideration of three bound components. The first of the components corresponds to the computation of the within-cluster sums of dissimilarities for the assigned objects. When p objects have been assigned to clusters, this component is computed as follows:

$$\text{Component 1} = \sum_{i=1}^{p-1} \sum_{j=i+1}^{p} \left(a_{ij} \mid \lambda_i = \lambda_j \right). \tag{4.1}$$

The second component corresponds to dissimilarity contributions between each of the unassigned objects and the objects that have already been assigned to groups. The contribution to the second component for each of the unassigned objects is reflected by the minimum possible contribution across all clusters. The component, which need not be obtained if there is an empty cluster, is computed as follows:

$$\text{Component 2} = \sum_{i=p+1}^{n} \min_{k=1,\ldots,K} \left[\sum_{j=1}^{p} \left(a_{ij} \mid \lambda_j = k \right) \right]. \tag{4.2}$$

The third component of the bound pertains to possible contributions to the objective function that stem from dissimilarities among the unassigned objects. The computation of this component uses (ascending) ranks of the pairwise dissimilarity values. We define ξ_e as the $e = 1,...,$ $[(n - p)(n - p - 1)/2]$ ascending rank-ordered dissimilarity values corresponding to pairs of objects from $p + 1,..., n$. The minimum number of possible dissimilarity terms among the unassigned objects that would be realized in a completed solution is given by:

$$\varphi = \alpha\beta + K\alpha(\alpha-1)/2, \qquad (4.3)$$

where α is the greatest integer that is less than or equal to $[(n - p) / K]$, and $\beta = \mathrm{mod}_K(n - p)$. The third component of the bound is obtained from summation of the φ smallest dissimilarities:

$$\text{Component 3} = \sum_{e=1}^{\varphi} \xi_e . \qquad (4.4)$$

The partial solution is pruned in Step 3 if:

$$\text{Component 1} + \text{Component 2} + \text{Component 3} \geq f_2(\lambda^*). \qquad (4.5)$$

As pointed out by Klein and Aronson (1991), the ranked dissimilarity bounds associated with Component 3 can be computed and stored prior to initiation of the branch-and-bound algorithm. Thus, the third bound component is actually computed in the INITIALIZE step *after* the matrix is reordered, and used as needed in Step 3. For every *Position*, Component 3 can be calculated once and stored for use for every partial evaluation of the last $n - Position$ objects. Using the previously employed notation (e.g., *num_k* is the number of desired clusters), the computational implementation of storing values for Component 3 is the final stage of the initialization procedure.

/* **Initialize** storage for Component3 for use in partial evaluations */
for *Position* = 1 to *n*
  Component3(*Position*) = 0
next *Position*
for *Position* = 1 to *n*
  Index = 0
  /* **Collect** dissimilarities of last *n – Position* objects */
  for i = *Position* + 1 to n – 1
    for j = i + 1 to n
      Index = Index + 1
      collection(Index) = **A**(i, j)

```
      next j
    next i
  /* Rank order the collection of dissimilarities */
  for i = 1 to Index – 1
    for j = i + 1 to Index
      if collection(j) < collection(i) then
        hold = collection(j), collection(j) = collection(i),
             collection(i) = hold
      end if
    next j
  next i
  /* Determine the smallest number of dissimilarities needed for
     Component 3 using equation (4.3) */
  Set beta = n – Position and alpha = 0
  while beta > num_k
    beta = beta – num_k
    alpha = alpha + 1
  loop
  Index = alpha * beta + num_k * alpha * (alpha – 1) / 2
  for i = 1 to Index
    Component3(Position) = Component3(Position) + collection(i)
  next i
next Position      /* End computation of Component3 bounds */
```

With the stored values for Component 3, the implementation of the partial solution evaluation is expedited. During any given evaluation of the partial solution, the complete bound is computed as follows:

```
Set Component1 = 0, Component2 = 0, Prune = False, Fathom = True.
/* Calculate Component 1 for assigned objects */
for i = 1 to Position – 1
  for j = i + 1 to Position
    if lambda(i) = lambda(j) then Component1 = Component1 + A(i, j)
  next j
next i
/* Calculate Component 2 for unassigned objects by the minimum
possible contribution to the objective function value across all clusters */
for unassigned = Position + 1 to n
  for k = 1 to num_k
    sum(k) = 0
    for assigned = 1 to Position
      if lambda(assigned) = k then
```

4.3 The PARTIAL SOLUTION EVALUATION Step

```
            sum(k) = sum(k) + A(unassigned, assigned)
          end if
        next assigned
      next k
      MinK = sum(1)
      for k = 2 to num_K
        if sum(k) < MinK then MinK = sum(k)
      next k
      Component2 = Component2 + MinK
    next unassigned
    /* Recall Component 3 value for the Position and determine acceptabil-
    ity of partial solution evaluation */
    if Component1 + Component2 + Component3(Position) ≥ incumbent
    then
      PART_EVAL = Prune
    else
      PART_EVAL = Fathom
    end if              /* End of partial solution evaluation */
```

Although the ranked dissimilarity bounds for Component 3 are advantageous from a computational standpoint, Component 3 of the bound is not particularly sharp (Brusco, 2003; Hansen & Jaumard, 1997). In other words, Component 3 often greatly understates possible dissimilarity contributions among the unassigned objects.

Brusco (2003) proposed replacing Component 3 with a bound, still calculated and stored in the initialization subroutine, that is obtained via an optimal K-cluster solution for the unassigned objects. The rationale is that the optimal solution for $n - p$ unassigned objects provides a much better indicator of what can be realized for these objects. Using this replacement for Component 3, the algorithm works back-to-front. If we are trying to partition a group of objects into K clusters, then the optimal value for the final K objects is zero because each object would be assigned to its own cluster. We then run the branch-and-bound algorithm for the final $K + 1$ objects, which calls upon the value previously found (zero) during the partial solution evaluation. Once solved we have our optimal value for Component3($n - (K + 1)$), we call upon the two previous optimal values to execute the branch-and-bound algorithm for the final $K + 2$ objects. The process continues until we consider all n objects when we will have $n - 1$ previous optimal values. As a result, significantly improved bound components will be available as the search moves deeper into the tree. The principle of using suborders of the data set to

50   4 Minimum Within-Cluster Sums of Dissimilarities Partitioning

develop improved bounds has also been recognized as useful when using criterion (2.3), and will be discussed in more detail in Chapter 5.

4.4 A Numerical Example

The branch-and-bound algorithm for minimizing the within-cluster sums of dissimilarity measures (2.2) can be applied to the 6×6 dissimilarity matrix shown in Table 3.1 in Chapter 3. For illustration purposes, we will assume that the INITIALIZE step uses the within-cluster sums of dissimilarities corresponding to the optimal minimum diameter 2-cluster partition {1, 2, 5, 6}, {3, 4}, which provides an initial upper bound for the within-cluster sums of dissimilarities of $f_2(\lambda^*) = 276$.

The computation of Component 3 for this data set is shown in Table 4.1, for various values of the position pointer, p. When $p \geq 4$, Component 3 is 0 because there are two or fewer unassigned objects, which could thus be assigned to different clusters and, therefore, produce no contribution to the objective function. When $p = 3$, objects 4, 5, and 6 are unassigned, but at least two of these three objects must be assigned to the same cluster because there are only $K = 2$ clusters. Thus, at least one of the terms $a_{45} = 37$, $a_{46} = 29$, or $a_{56} = 34$ must be picked up for the criterion contribution, and the minimum of these terms is the appropriate value for Component 3. For $p = 2$, objects 3, 4, 5, and 6 are unassigned, and the minimum number of possible terms collected among these four objects in the criterion function is two. Thus, when $p = 2$, the Component 3 bound is the sum of the minimum two elements among ($a_{34} = 43$, $a_{35} = 63$, $a_{36} = 53$, $a_{45} = 37$, $a_{46} = 29$, $a_{56} = 34$), which is $29 + 34 = 63$. Finally, when $p = 1$, objects 2 through 6 are unassigned. The minimum number of collected terms will occur when three of the $n - p = 6 - 1 = 5$ unassigned objects are placed in one cluster and two of the unassigned objects in the other cluster. Such an allocation will yield four terms for Component 3. For $p = 1$, the Component 3 bound is the sum of the minimum four elements among ($a_{23} = 17$, $a_{24} = 88$, $a_{25} = 50$, $a_{26} = 41$, $a_{34} = 43$, $a_{35} = 63$, $a_{36} = 53$, $a_{45} = 37$, $a_{46} = 29$, $a_{56} = 34$), which is $17 + 29 + 34 + 37 = 117$.

After initializing the incumbent solution at 276 with a heuristic procedure, the iterations of the branch-and-bound algorithm for a 2-cluster partition based on (2.2) are shown in Table 4.2. The first two columns provide the row label and the cluster assignments for each object. These are followed by columns for the pointer, p, the within-cluster dissimilarity sums of clusters $k = 1$ and $k = 2$ (whose sum is bound Component 1), bound Components 2 and 3, and the dispensation of the partial solution.

Table 4.1. Component 3 bounds for data in Table 3.1.

Pointer Position	Ranked Dissimilarities	Component 3
$p = 1$	17, 29, 34, 37, 41, 43, 50, 53, 63, 88	117
$p = 2$	29, 34, 37, 43, 53, 63	63
$p = 3$	29, 34, 37	29
$p \geq 4$		0

The branch-and-bound solution for (2.2) required more iterations than the corresponding solution for (2.1), as indicated by the greater number of rows in Table 4.2 relative to Table 3.2 (18 versus 11). The algorithm proceeds rather steadily through the tree, finding a new incumbent solution in row 9. This solution has a within-cluster sum of dissimilarities value of $f_2(\lambda^*) = 211$, and corresponds to the partition $\{1, 2, 3\}, \{4, 5, 6\}$. An interesting note is that this partition, which is subsequently verified as optimal, is appreciably different from the optimal minimum diameter partition obtained in section 3.4.

One of the most beneficial pruning operations occurs in row 10 of Table 4.2. The partial solution associated with this row shows an assignment of objects 1 and 2 to cluster 1 and object 3 to cluster 2 (thus, three positions have been assigned; so, $p = 3$). Cluster 1, therefore, produces the within-cluster dissimilarity value of $a_{12} = 35$, whereas cluster 2 does not yield any direct contribution to the bound. Thus, the total contribution from Component 1 is 35. Component 2 is computed by evaluating the minimum possible contributions of objects 4, 5, and 6 given tentative assignments to either cluster 1 or cluster 2. If object 4 is assigned to cluster 1, then the resulting contribution would be $a_{14} + a_{24} = 140$; however, assignment to cluster 2 would yield only $a_{34} = 43$ and this smaller value is object 4's contribution to Component 2 of the bound. Using similar logic, the bound contributions of objects 5 and 6 are $a_{35} = 63$ and $a_{36} = 53$, respectively. Thus, the total bound contribution for Component 2 is $43 + 53 + 63 = 159$.

Table 4.2. Branch-and-bound summary for the data in Table 3.1.

Row	λ	p	Sums k=1	Sums k=2	Components 2	Components 3	Dispensation
1	1	1	0	0	0	117	Branch forward
2	1 1	2	35	0	0	63	Branch forward
3	1 1 1	3	111	0	0	29	Branch forward
4	1 1 1 1	4	291	0	0	0	Prune (291 ≥ 274), Branch right
5	1 1 1 2	4	111	0	66	0	Branch forward
6	1 1 1 2 1	5	248	0	29	0	Prune (277 ≥ 274), Branch right
7	1 1 1 2 2	5	111	37	63	0	Branch forward
8	1 1 1 2 2 1	6	252	37	0	0	Prune (289 ≥ 276), Branch right
9	1 1 1 2 2 2	6	111	100	0	0	*New Incumbent, $f_2(\lambda^*) = 211$
10	1 1 2	3	35	0	159	29	Prune (223 ≥ 211), Retract
11	1 2	2	0	0	134	63	Branch forward
12	1 2 1	3	59	0	179	29	Prune (267 ≥ 211), Branch right
13	1 2 2	3	0	17	123	29	Branch forward
14	1 2 2 1	4	52	17	137	0	Branch forward
15	1 2 2 1 1	5	113	17	94	0	Prune (224 ≥ 211), Branch right
16	1 2 2 1 2	5	52	130	76	0	Prune (258 ≥ 211), Retract
17	1 2 2 2	4	0	145	71	0	Prune (216 ≥ 211), Retract
18	2	1					TERMINATE

The third component of the bound is Component 3 = 29, which corresponds to the appropriate contribution from this component when $p = 3$ (as shown in Table 4.1). Using the determined values of Component 1 = 35, Component 2 = 159, and Component 3 = 29, we have: (35 + 159 + 29) = 223 ≥ $f_2(\lambda^*)$ = 211, which results in the pruning of the partial solution. Notice that the absence of any one of the three bound components would lead to a failure to prune the partial solution.

As described in section 4.2, the third component of the bound can be improved using an optimal solution for the unassigned objects. To illustrate, consider row 11 from Table 4.2. The partial solution for this row corresponds to the assignment of $p = 2$ objects: object 1 to cluster 1 and object 2 to cluster 2. The bound components for this partial solution are Component 1 = 0, Component 2 = 134, and Component 3 = 63. Because $(0 + 134 + 63) = 197 < f_2(\lambda^*) = 211$, it is necessary to pursue the corresponding branch and plunge deeper into the tree. However, if we obtain the optimal 2-cluster partition for the four objects (3, 4, 5, and 6), the resulting partition is {3, 4} {5, 6}, which yields an objective value of $a_{34} + a_{56} = 43 + 34 = 77$. This value of 77 is a better reflection of the best that can be achieved among the last four objects, and can therefore replace the value of 63 for Component 3. As a result, the new bound evaluation is $(0 + 134 + 77) = 211 \geq f_2(\lambda^*) = 211$. Using the improved bound component, it is clear that the current partial solution cannot be pursued so as to provide a partition that yields an objective criterion value better than the incumbent value of 211. The partial solution is pruned at row 11 and retraction occurs, precluding the evaluation of the solutions in rows 12 through 17 and resulting in algorithm termination and return of the optimal solution. Again, we reiterate that this solution procedure requires a back-to-front approach using successive executions of the branch-and-bound algorithm. The main algorithm is executed for the last $n - (K + 1)$ objects and the optimal objective function value is stored as Component$3(n - (K + 1))$ for use in the execution for the last $n - (K + 2)$ objects and so forth.

4.5 Application to a Larger Data Set

To facilitate comparison to the 3-cluster results for (2.1) in section 3.5, we applied the branch-and-bound algorithm for (2.2) to the lipread consonant dissimilarity data in Table 3.3 under the assumption of $K = 3$ clusters. The within-cluster submatrices are shown in Table 4.3. The cluster sizes for the optimal partition based on (2.2) are more equally balanced than the minimum-diameter solution in Chapter 3. This is expected based on the smaller number of pairwise dissimilarity elements in the sum when cluster sizes are approximately equal.

Cluster 1 in Table 4.3 consists exclusively of letters with the "long e" sound, whereas cluster 2 contains the letters with the "eh" sound. Cluster 3 is more of a mixed bag of letters. The three clusters provide roughly equal contributions to the within-cluster sum of pairwise dissimilarities. The within-cluster sum for cluster 1 is $(141 + 176 + 118 + ... + 180 + 147 + 204) = 4391$, and the corresponding contributions from clusters 2 and 3 are 4467 and 4319, respectively. The total within cluster sum of pairwise dissimilarity elements for the three-cluster partition is $f_2^*(\pi_3) = 4391 + 4467 + 4319 = 13177$.

Clusters 1 and 2 in Table 4.3 are rather similar to clusters 1 and 2 for the ($K = 3$) minimum-diameter partition in Table 3.5. Clusters 1 and 2 in Table 4.3 are compact clusters, as described in Chapter 2, with diameters of 275 and 276, respectively. Cluster 3 in Table 4.3, however, is not a compact cluster. The diameter for this cluster is 363 (the dissimilarity measure for objects pair (r, y)). There are also several other large pairwise dissimilarity elements in cluster 3 (349, 337, 325, 324), which suggests that this cluster does not consist of homogeneous letters. If object j is moved from cluster 3 to cluster 1 and objects k and r are moved from cluster 3 to cluster 2, then the diameter of cluster 3 greatly improves, but the within-cluster sums for clusters 1 and 2 increase markedly with the larger number of objects in those clusters.

4.6 Strengths and Limitations of the Within-Cluster Sums Criterion

As we have seen in the examples of this section, one of the limitations of the within-cluster sum of pairwise dissimilarities index is a tendency to produce clusters of approximately the same size. This occurs because the number of pairwise dissimilarity terms collected in the computation is smaller when clusters are roughly the same size. For example, if $n = 20$ objects are partitioned into $K = 4$ groups of size five, then there are 4(5(5 − 1)/2) = 40 pairwise dissimilarities that are summed to compute the index. However, if the 20 objects are partitioned into three groups of size one and one group of size 17, then there are 17(16)/2 = 136 pairwise dissimilarity sums used in the index computation. The potential adverse result from this characteristic is that, in an effort to reduce the within-cluster sums of pairwise dissimilarities by balancing cluster sizes, one or more of the clusters might not be compact. If total interaction is of paramount concern, not compact clusters, then this might not be problematic. On the other hand, if homogeneity of all clusters is important, then some integration of (2.1) and (2.2) in the objective function is advisable. This suggests the implementation of a bicriterion partitioning approach, which is discussed in Chapter 6.

4.6 Strengths and Limitations of the Within-Cluster Sums Criterion

Table 4.3. Within-cluster submatrices for lipread consonants data obtained when optimizing (2.2).

Cluster 1	b	c	d	g	p	t	v	z
b	--							
c	141	--						
d	176	118	--					
g	155	149	147	--				
p	69	214	149	275	--			
t	182	122	90	169	176	--		
v	227	182	173	173	200	131	--	
z	129	49	82	182	180	147	204	--

Cluster 2	f	h	l	m	n	s	x
f	--						
h	271	--					
l	249	233	--				
m	216	267	175	--			
n	243	200	149	243	--		
s	204	249	147	276	176	--	
x	255	178	176	271	214	75	--

Cluster 3	j	k	q	r	w	y
j	--					
k	182	--				
q	300	247	--			
r	324	214	192	--		
w	349	318	298	337	--	
y	286	325	284	363	300	--

The *raw* within-cluster sum of pairwise dissimilarities is somewhat overshadowed by the standardized (by dividing by cluster size) variation discussed in the next chapter. This is attributable to the relationship between the standardized within-cluster sums of pairwise dissimilarities and the minimization of the within-cluster sum of squared deviation from cluster centroids when **A** is comprised of squared Euclidean distances between pairs of objects. Nevertheless, the raw sum is a worthy alternative when **A** is not defined in this manner.

4.7 Available Software

We have made available a software program for minimizing the within-cluster sum of dissimilarities, which can be accessed and downloaded from either http://www.psiheart.net/quantpsych/monograph.html or http://acns.fsu.edu/~mbrusco. The program *bbwcsum.for* is similar to the algorithm for minimizing (2.2) as described by Brusco (2003). The algorithm does not use the bounds suggested by Klein and Aronson (1991), but instead employs a staged solution process. Specifically, (2.2) is minimized for $K + 2, K + 3, K + 4, ..., n$ objects. As described in section 4.2, this process considerably strengthens the third component of the bounding process. In fact, for any particular stage each of the optimal objective values obtained from previous stages are incorporated in the bounding process.

The file structure for *bbwcsum.for* is identical to those of *bbdiam.for* and *bbdisum.for* as described in section 3.8. The only input required from the user is the form of the matrix (half-matrix or full matrix) and the number of clusters. However, *bbwcsum.for* is far more restricted with respect to the sizes of clustering problems that can be successfully handled (notice that the prompt for the number of clusters is limited to 10 and, more practically, a limit of 5 or 6 might be appropriate). Despite the incorporation of information from previous stages of the solution process, the bounds for (2.2) are not especially sharp relative to those for (2.1) or even (2.3), and optimal solution using *bbwcsum.for* is only practical for up to about 50 objects and 5 or 6 clusters. Optimal solutions might be possible for slightly larger problems if clusters are well separated, and impossible for some smaller problem instances where separation is poor.

To illustrate the operation of *bbwcsum.for*, we present the following screen display information for execution of the program for the data in Table 3.3 at $K = 3$ clusters.

The user input information is:

```
> TYPE 1 FOR HALF MATRIX OR TYPE 2 FOR FULL MATRIX
> 1
> PLEASE INPUT NUMBER OF CLUSTERS 2 TO 10
> 3
```

The solution output is:
```
NUMBER OF OBJECTS    4 Z =      239.00000************
NUMBER OF OBJECTS    5 Z =      502.00000    549.00010
NUMBER OF OBJECTS    6 Z =      670.00000    780.00010
NUMBER OF OBJECTS    7 Z =      857.00000    953.00010
NUMBER OF OBJECTS    8 Z =     1384.00000   1384.00010
NUMBER OF OBJECTS    9 Z =     1966.00000   1966.00010
NUMBER OF OBJECTS   10 Z =     2522.00000   2522.00010
NUMBER OF OBJECTS   11 Z =     3155.00000   3155.00010
NUMBER OF OBJECTS   12 Z =     4067.00000   4118.00010
NUMBER OF OBJECTS   13 Z =     4714.00000   4714.00010
NUMBER OF OBJECTS   14 Z =     5794.00000   5798.00010
NUMBER OF OBJECTS   15 Z =     6944.00000   7053.00010
NUMBER OF OBJECTS   16 Z =     8071.00000   8071.00010
NUMBER OF OBJECTS   17 Z =     9027.00000   9027.00010
NUMBER OF OBJECTS   18 Z =    10465.00000  10465.00010
NUMBER OF OBJECTS   19 Z =    11264.00000  11341.00010
NUMBER OF OBJECTS   20 Z =    12098.00000  12098.00010
NUMBER OF OBJECTS   21 Z =    13177.00000  13177.00010

  MINIMUM WITHIN CLUSTER SUMS OF DISSIMILARITIES
       13177.00000
  TOTAL CPU TIME =              4.86
  1  1  1  2  1  2  3  3  2  2  2  1  3  3  2  1
  1  3  2  3  1

  Stop - Program terminated.
```

We have included the output for each stage of the solution process. The solution output for each stage provides the optimum solution values at each stage and, immediately to the right of these numbers, the upper bound for that particular stage. Although this information can easily be "commented out" from the main program, we believe that it is helpful to the user because it enables progress to be tracked. For example, if a user were attempting to solve a problem with $n = 50$ objects in $K = 6$ clusters and the solution process was proceeding very slowly when the number of objects was only in the mid-20s, then we would be wise to abandon the solution attempt because the time required to solve the full problem is apt to be astronomical. Fortunately, for the data in Table 3.3, the branch-and-bound algorithm provides the optimal solution in approximately 2.5 seconds (Pentium IV PC, 2.2 GHz). The optimal solution corresponds to the results reported in Table 4.3 and discussed in section 4.4.

5 Minimum Within-Cluster Sums of Squares Partitioning

5.1 The Relevance of Criterion (2.3)

As pointed out in sections 2.1 and 4.5, criterion (2.3) has a particularly important role in cluster analysis. We refer to (2.3) as the *standardized* within-cluster sums of dissimilarities, where standardization occurs via the division of the within-cluster sums by the number of objects assigned to the cluster. When the elements of **A** correspond to squared Euclidean distances between pairs of objects, then an optimal solution to (2.3) minimizes the within-cluster sums of squared deviations between objects and their cluster centroids. That is, minimizing the sum of the standardized within-cluster sums of dissimilarities is equivalent to minimizing the within-cluster sums of squares. This is the well-known K-means criterion (Forgy, 1965; MacQueen, 1967), for which heuristic programs are available with most commercial statistical software packages.

Although (2.3) is a viable criterion when the elements of **A** do not correspond to squared Euclidean distances between pairs of objects, greater caution must be taken in branch-and-bound implementation for such instances. Because the dissimilarities are assumed to be nonnegative, the addition of an object to an existing cluster cannot decrease the *raw* within-cluster sum of pairwise dissimilarities; however, the *standardized* within-cluster sum can decrease. Consider, for example, a hypothetical cluster with two objects $\{1, 2\}$. Suppose that $a_{12} = 5$, and, thus, the raw sum of dissimilarities for the cluster is 5 and the standardized sum is 5/2 = 2.5 (divide by two because there are two objects in the cluster). Suppose we wish to add object 3 to the cluster and $a_{13} = a_{23} = 1$. The raw sum for the cluster $\{1, 2, 3\}$ increases to $5 + 1 + 1 = 7$, but the standardized sum decreases to 7/3 = 2.33.

The possibility for reduction of the standardized within-cluster sums of pairwise dissimilarities when an object is added to a cluster precludes the use of some of the bound components available for (2.2). In other words, the development of strong bounds (analogous to Components 2 and 3 described in section 4.2) for (2.3) are not readily available for general dissimilarity matrices. This seriously diminishes the effectiveness of branch-and-bound for (2.3) when **A** does not have the necessary metric properties.

Stronger bound components are available, however, when considering the special case of **A** corresponding to squared Euclidean distances. Under such conditions, Koontz et al. (1975) proved that the addition of an object to a cluster cannot possibly decrease the standardized within-cluster sums of dissimilarities. In fact, an increase in the standardized sum will not occur only if the added object is positioned exactly at the centroid of the cluster (in which case, the change in the standardized sum is 0). Throughout the remainder of this chapter, we will assume that **A** is a matrix of squared Euclidean distances.

Notice that the implementation is extremely similar to that of the implementation of criterion (2.2) in Chapter 4.

```
/* Compute the within-cluster sum of dissimilarities */
for k = 1 to num_k           /* Initialize sum/centroid for each cluster */
   Centroid(k) = 0
next k
for Position1 = 1 to n – 1
   k1 = lambda(Position1)
   for Position2 = Position1 + 1 to n
      k2 = lambda(Position2)
      if k1 = k2 then
         Centroid(k1) = Centroid(k1) + A(Position1, Position2)
      end if
   next Position2
next Position1
EVALUATION = 0
for k = 1 to num_k
   EVALUATION = EVALUATION + Centroid(k) / CSize(k)
next k                       /* End computation of criterion (2.3) */
```

5.2 The INITIALIZE Step

We recommend a traditional K-means heuristic (Forgy, 1965; MacQueen, 1967) and/or an exchange heuristic to provide a good initial upper bound for (2.3). K-means heuristics begin with an initial set of cluster centroids or "seed points," which correspond to variable means for each

5.2 The INITIALIZE Step

cluster. Objects are assigned to their nearest cluster, and the centroids are recomputed. The iterative process of reassignment and recomputation continues until no objects change cluster assignment on a particular iteration. Although the K-means algorithm is extremely efficient, the algorithm is sensitive to the initial seed points and, therefore, multiple restarts of the algorithm using different seed points are recommended. The importance of using multiple seed points has recently been demonstrated in a study by Steinley (2003).

For our demonstrations of the branch-and-bound process, we assume a matrix of squared Euclidean distances. We have applied the exchange algorithm described in Chapters 3 and 4 after appropriate modification for criterion (2.3) to a randomly generated solution to produce the initial incumbent solution in our demonstrations. Given the similarity between criterion (2.2) and criterion (2.3), the INITIALIZE step of Chapter 4 is naturally very similar to the INITIALIZE step for this chapter. However, for criterion (2.3) we need not compute Component 3, but we do need to track the cluster sizes and ensure that no feasible solution contains an empty cluster. The need to track the cluster sizes is a byproduct of the possibility that the standardized within-cluster sum of squares will decrease when a new object is added. Although this initialization works well, a tandem approach, using K-means followed by the exchange algorithm, would quite possibly yield better results. The incumbent solution will be noted as $f_3(\lambda^*)$ in the remainder of the chapter.

```
k = 0
for c = 1 to num_k
   CSize(c) = 0
next c
for i = 1 to n
   k = k + 1
   if k > num_k then k = 1
   lambda(i) = k
   CSize(k) = CSize(k) + 1
next i
flag = True
while flag
   Set flag = False, flag1 = True, and tempflag = 1
   /* SINGLE OBJECT RELOCATION */
   while flag1 and (tempflag < 10)
      tempflag = tempflag + 1
      flag1 = False
```

```
for c = 1 to num_k   /* Find sum of dissimilairities for each cluster */
  WithinSum(c) = 0
next c
for cluster = 1 to num_k
  for object = 1 to n
    if lambda(object) = cluster then
      for object2 = object + 1 to n
        if lambda(object2) = cluster then
          WithinSum(cluster) = WithinSum(cluster)
                                          + A(object, object2)
      next object2
    end if
  next object
next cluster
/* Determine effect on objective function of moving each object
    to other clusters */
for object = 1 to n
  for NewCluster = 1 to num_k
    if NewCluster <> lambda(object) then
      NewWithin = WithinSum(lambda(object))
      for bedfellow = 1 to n
        if lambda(object) = lambda(bedfellow) then
          NewWithin = NewWithin - A(object, bedfellow)
        end if
      next bedfellow
      NewWithin2 = WithinSum(NewCluster)
      for object2 = 1 to n
        if lambda(object2) = NewCluster then
          NewWithin2 = NewWithin2 + A(object, object2)
        end if
      next object2
      size1 = CSize(lambda(object)) and size2 = CSize(NewCluster)
      if (WithinSum(lambda(object)) / size1
          + WithinSum(NewCluster) / size2)
        > (NewWithin / (size1 - 1) + NewWithin2 / (size2 + 1)) then
        /* Reassign */
        CSize(NewCluster) = CSize(NewCluster) + 1
        CSize(lambda(object)) = CSize(lambda(object)) − 1
        WithinSum(lambda(object)) = NewWithin
        WithinSum(NewCluster) = NewWithin2
        lambda(object) = NewCluster
```

```
                flag1 = True
              end if
            end if
          next NewCluster
        next object
      loop flag1
      Set flag2 = True
      while flag2            /* PAIRWISE INTERCHANGE */
        flag2 = False
        for i = 1 to n - 1
          for j = i + 1 to n
            if lambda(i) <> lambda(j) then
              WithinI = WithinSum(lambda(i))
              WithinJ = WithinSum(lambda(j))
              for bedfellow = 1 to n
                if lambda(bedfellow) = lambda(i) then
                  WithinI = WithinI - A(i, bedfellow) + A(j, bedfellow)
                end if
                if lambda(bedfellow) = lambda(j) then
                  WithinJ = WithinJ - A(j, bedfellow) + A(i, bedfellow)
                end if
              next bedfellow
              size1 = CSize(lambda(i)) and size2 = CSize(lambda(j))
              if (WithinSum(lambda(i)) / size1
                              + WithinSum(lambda(j)) / size2)
                          > (WithinI / size1 + WithinJ / size2) then
                WithinSum(lambda(i)) = WithinI
                WithinSum(lambda(j)) = WithinJ
                swap = lambda(i), lambda(i) = lambda(j), lambda(j) = swap
                flag2 = True
              end if
            end if
          next j
        next i
      loop                   /* End of pairwise interchange loop; flag2 */
    loop                     /* End of exchange algorithm; flag */
```

5.3 The PARTIAL SOLUTION EVALUATION Step

Following Koontz et al. (1975) and Diehr (1985), the evaluation of partial solutions initially focuses on two bound components. The first of the components is the sum of the standardized within-cluster sums of dissimilarities for the p assigned objects:

$$\text{Component 1} = \sum_{k=1}^{K} \left[\frac{\sum_{i=1}^{p-1} \sum_{j=i+1}^{p} (a_{ij} | \lambda_i = \lambda_j = k)}{n_k} \right]. \tag{5.1}$$

The second bound component, which is very weak, examines the effect of adding each of the unassigned objects to each of the clusters. For each of the unassigned objects, the minimum possible contribution across all clusters is selected, and the maximum of these minimums is selected as the second bound component. More formally:

$$\text{Component 2} = \max_{i=p+1,\ldots,n} \left[\min_{k=1,\ldots,K} (\Delta_i(C_k)) \right], \tag{5.2}$$

where $\Delta_i(C_k)$ is the increase in the standardized within-cluster sum of pairwise dissimilarities that will occur from the inclusion of object i in cluster C_k. A partial solution is pruned if:

$$\text{Component 1} + \text{Component 2} \geq f_3(\lambda^*). \tag{5.3}$$

Set Component1 = 0, Component2 = 0, prune = False, fathom = True, and PART_EVAL = fathom.
/* **Calculate** Component 1 for assigned objects */
for c = 1 to *num_k*
  for i = 1 to *Position* − 1
    for j = i + 1 to *Position*
    if lambda(i) = lambda(j) then Component1 = Component1 + **A**(i, j)
    next j
  next i
next c
/* **Calculate** Component 2 for unassigned objects by the minimum possible contribution to the objective function value across all clusters. */
for unassigned = *Position* + 1 to *n*
  for c = 1 to *num_k*

```
    Sum(c) = 0
    for assigned = 1 to Position
      if lambda(assigned) = c then
        Sum(c) = Sum(c) + A(unassigned, assigned)
    next assigned
  next c
  MinK = Sum(1)
  for c = 2 to num_k
    if Sum(c) < MinK then MinK = Sum(c)
  next c
  if MinK > Component2 then Component2 = MinK
next unassigned
/* Determine acceptability of partial solution evaluation */
if Component1 + Component2 >= incumbent then
  PART_EVAL = prune
else
  PART_EVAL = fathom
end if           /* End of partial solution evaluation */
```

Because the second component of the bound is weak, a better option is to replace Component 2 with an optimal solution for the unassigned objects. Again, the motivation here is that the branch-and-bound algorithm using Components 1 and 2 is effective for small values of n, and thus can be used to rapidly obtain optimal solutions for suborders of the set of objects. This is similar to the approach outlined in section 4.2. Koontz et al. (1975) and Diehr (1985) went a step further and discussed the initial partitioning of the object set, S, into e manageably sized subsets $\{S_1, S_2, ..., S_e\}$. Because the sizes for the subsets are small, they can be optimally solved using the bounding components associated with (5.1) and (5.2). The optimal criterion values for the object subsets 1, 2,..., e are denoted $f_3^*(S_1), f_3^*(S_2), ..., f_3^*(S_e)$, respectively. The optimal solutions for the subsets are subsequently pieced together in a systematic fashion until an optimal solution for the complete object set is provided. This decomposition approach is based on bound improvement, which is achievable because the optimal criterion value solution for the complete object set must equal or exceed the sum of the optimal objective criterion values for the subsets. Specifically:

$$f_3^*(S) \geq f_3^*(S_1) + f_3^*(S_2) + ... + f_3^*(S_e). \qquad (5.4)$$

As a simple example, suppose that $S_1 = \{1, 2,..., p\}$ and $S_2 = \{p + 1, p + 2, ..., n\}$. If the optimal solution for S_2 is available, then a partial sequence of p objects can be pruned if:

$$\text{Component } 1 + f_3^*(S_2) \geq f_3(\lambda^*). \tag{5.5}$$

The rationale here is that, even if Component 1 represented the optimal assignment for objects $1,...,p$, the optimal criterion value for the complete set would have to be at least as large as the left side of (5.5). Thus, if the left-hand-side of (5.5) exceeds the incumbent criterion value, there is no possible way that the current partial solution could ultimately lead to a better solution.

5.4 A Numerical Example

To illustrate the application of the branch-and-bound algorithm to minimize the standardized within-cluster sums, we consider $n = 6$ objects in two-dimensional space, as depicted in Figure 5.1. The 6×6 matrix of integral squared Euclidean distances between points in Figure 5.1 is shown in Table 5.1. We will assume that the INITIALIZE step has employed a heuristic that yields an initial upper bound for the standardized within-cluster sums of pairwise dissimilarities (2.3) of $f_3(\lambda^*) = 23.1$. Further, for the purposes of our demonstration, $S_2 = \{4, 5, 6\}$. In this particular problem, the optimal partition for $K = 2$ is $\{4, 6\}$ $\{5\}$, which yields $f_3^*(S_2) = a_{12}/2 = 20/2 = 10$. Whenever the pointer position is $p \leq 3$, we will use this bound (denoted Component 3), otherwise we will use Component 2 from (5.2).

The iterations of the branch-and-bound algorithm for a 2-cluster partition based on (2.3) are shown in Table 5.2. A particularly crucial pruning of the tree occurs in row 2. For this partial solution, both objects 1 and 2 are assigned to cluster 1 and thus Component 1 of the bound is $a_{12} = 29/2 = 14.5$. Adding this component to $f_3^*(S_2) = 10$ results in $24.5 \geq f_3(\lambda^*) = 23.1$, and the branch is pruned. Without the exceptionally strong bound from the optimal solution for S_2, this pruning would not have been possible. Table 5.2 also shows that Component 2 can, in some instances, also facilitate pruning of the tree (e.g., row 18 and row 20); however, most of these instances do not occur until the pointer has moved rather deep into the tree.

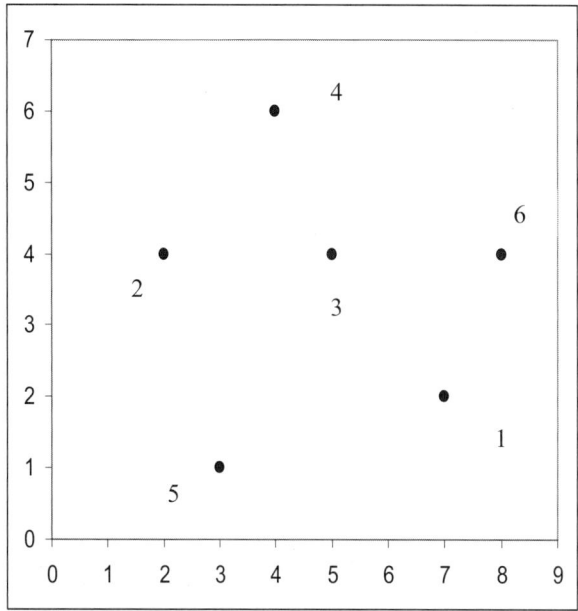

Figure 5.1. Six objects in a two-dimensional space.

Table 5.1. Example dissimilarity matrix of squared Euclidean distances between points in Figure 5.1.

	1	2	3	4	5	6
1	----					
2	29	----				
3	8	9	----			
4	25	8	5	----		
5	17	10	13	26	----	
6	5	36	9	20	34	----

The optimal partition for the data set in Table 5.1 (derived from Figure 5.1) using criterion (2.3) is {1, 6}, {2, 3, 4, 5}, which yields a standardized within-cluster sums of dissimilarities value of 20.25. For comparative purposes, considering the optimal branch-and-bound solutions for the same data set using (2.1) and (2.2) is illustrative. The branch-and-bound tables for minimum-diameter partitioning (2.1) and raw within-cluster sums of dissimilarities (2.2) are shown in Tables 5.3 and 5.4, respectively.

Table 5.2. Branch-and-bound summary for criterion (2.3) for the data in Table 5.1

Row	λ	p	Standardized sums $k=1$	Standardized sums $k=2$	Components 2	Components 3	Dispensation
1	1	1	0	0	0	10	Branch forward
2	1 1	2	14.50	0	0	10	Prune (24.5 ≥ 23.1), Branch right
3	1 2	2	0	0	0	10	Branch forward
4	1 2 1	3	4.00	0	0	10	Branch forward
5	1 2 1 1	4	12.67	0	5.33	0	Branch forward
6	1 2 1 1 1	5	23.50	0	8.90	0	Prune (32.4 ≥ 23.1), Branch right
7	1 2 1 1 2	5	12.67	5.0	5.33	0	Branch forward
8	1 2 1 1 2 1	6	18.00	5.0	0	0	*New Incumbent, $f_3(\lambda^*) = 23$
9	1 2 1 1 2 2	6	12.67	26.67	0	0	Suboptimal, 39.3 ≥ 23, Retract
10	1 2 1 2	4	4.00	4.00	8.67	0	Branch forward
11	1 2 1 2 1	5	12.67	4.00	8.83	0	Prune (25.5 ≥ 23.1), Branch right
12	1 2 1 2 2	5	4.00	14.67	3.33	0	Branch forward
13	1 2 1 2 2 1	6	7.33	14.67	0	0	*New Incumbent, $f_3(\lambda^*) = 22$
14	1 2 1 2 2 2	6	4.00	33.50	0	0	Suboptimal, 37.5 ≥ 22, Retract
15	1 2 2	3	0	4.50	0	10	Branch forward
16	1 2 2 1	4	12.50	4.50	6.17	0	Prune (23.17 ≥ 22), Branch right
17	1 2 2 2	4	0	7.33	8.50	0	Branch forward
18	1 2 2 2 1	5	8.50	7.33	10.17	0	Prune (26.5 ≥ 22), Branch right
19	1 2 2 2 2	5	0	17.75	2.50	0	Branch forward
20	1 2 2 2 2 1	6	2.50	17.75	0	0	*New Incumbent, $f^*(\lambda^*) = 20.25$
21	1 2 2 2 2 2	6	0	34.00	0	0	Suboptimal, 34 ≥ 20.25, Retract
22	2	1					TERMINATE

The minimum-diameter partitioning algorithm was applied using an initial bound of 26, whereas the within-cluster sums of dissimilarities algorithm used a bound of 74. The optimal 2-cluster minimum-diameter partition is {1, 3, 4, 6}, {2, 5}, and the optimal 2-cluster raw within-cluster sum-of dissimilarities partition is {1, 3, 6}, {2, 4, 5}. Thus, three different criteria, (2.1), (2.2), and (2.3), yield three different optimal partitions for this small data set.

Table 5.3. Branch-and-bound summary for criterion (2.1) for the data in Table 5.1.

			Diameters		
Row	λ	p	$k=1$	$k=2$	Dispensation
1	1	1	0	0	Branch forward
2	1 1	2	29	0	Prune ($29 \geq 26$), Branch right
3	1 2	2	0	0	Branch forward
4	1 2 1	3	8	0	Branch forward
5	1 2 1 1	4	25	0	Branch forward
6	1 2 1 1 1	5	26	0	Prune (Component 2), Branch right
7	1 2 1 1 2	5	25	10	Branch forward
8	1 2 1 1 2 1	6	25	10	*New Incumbent, $f_1(\lambda^*) = 25$
9	1 2 1 1 2 2	6	25	34	Suboptimal, $34 \geq 25$, Retract
10	1 2 1 2	4	8	8	Branch forward
11	1 2 1 2 1	5	17	8	Prune (Component 2), Branch right
12	1 2 1 2 2	5	8	26	Prune ($26 \geq 25$), Retract
13	1 2 2	3	0	9	Branch forward
14	1 2 2 1	4	25	9	Prune ($25 \geq 25$), Branch right
15	1 2 2 2	4	0	9	Branch forward
16	1 2 2 2 1	5	17	9	Prune (Component 2), Branch right
17	1 2 2 2 2	5	0	26	Prune ($26 \geq 25$), Retract
18	2	1	0	0	TERMINATE

Table 5.4. Branch-and-bound summary for criterion (2.2) for the data in Table 5.1.

Row	λ	p	Sums		Components		Dispensation
			$k=1$	$k=2$	2	3	
1	1	1	0	0	0	31	Branch forward
2	1 1	2	29	0	0	20	Branch forward
3	1 1 1	3	46	0	0	20	Branch forward
4	1 1 1 1	4	84	0	0	0	Prune (84 ≥ 74), Branch right
5	1 1 1 2	4	46	0	46	0	Prune (92 ≥ 74), Retract
6	1 1 2	3	29	0	27	20	Prune (76 ≥ 74), Retract
7	1 2 1	3	8	0	32	20	Branch forward
8	1 2 1 1	4	38	0	44	0	Prune (82 ≥ 74), Branch right
9	1 2 1 2	4	8	8	44	0	Branch forward
10	1 2 1 2 1	5	38	8	48	0	Prune (94 ≥ 74), Branch right
11	1 2 1 2 2	5	8	44	14	0	Branch forward
12	1 2 1 2 2 1	6	22	44	0	0	*New Incumbent, $f_2(\lambda^*) = 66$
13	1 2 1 2 2 2	6	8	90	0	0	Suboptimal (98 ≥ 66), Retract
14	1 2 2	3	0	9	35	20	Branch forward
15	1 2 2 1	4	25	9	48	0	Prune (82 ≥ 66), Branch right
16	1 2 2 2	4	0	22	22	0	Branch forward
17	1 2 2 2 1	5	17	22	35	0	Prune (74 ≥ 66), Branch right
18	1 2 2 2 2	5	0	71	5	0	Prune (76 ≥ 66), Retract
19	2	1					TERMINATE

5.5 Application to a Larger Data Set

For our second demonstration using criterion (2.3), we use coordinates for 22 German towns, as reported by Späth (1980, p. 43). The two-dimensional coordinates for the 22 towns are provided in Table 5.5 and computational results for $2 \leq K \leq 8$ are summarized in Table 5.6. Table

5.6 provides the minimum standardized within-cluster sums of dissimilarities for each value of K, as well as a "standardized sums reduction ratio," $ssrr(K)$, which is analogous to the diameter reduction ratio described in section 3.5. The ratio represents the percentage reduction in criterion (2.3) that is realized from increasing the number of clusters from $K - 1$ to K. The $ssrr(K)$ column in Table 5.6 indicates that increasing the number of clusters from 3 to 4 provides a 44.87% reduction in criterion (2.3), whereas increasing from 4 to 5 clusters provides only a further 23.60% improvement.

Application of the branch-and-bound algorithm for minimum-diameter partitioning (2.1) to the data in Table 5.5 (assuming $K = 4$ clusters) yields a similar four cluster partition. The only difference is that town #19, Würzburg, is moved from the second cluster to the fourth cluster. Thus, for this particular data set, there is strong similarity between the four cluster partitions produced by (2.1) and (2.3).

Table 5.5. Coordinates for 22 German towns from Späth (1980, p. 43).

Town #	Name	x-axis	y-axis	Town #	Name	x-axis	y-axis
1	Aachen	-57	28	12	Köln	-38	35
2	Ausburg	54	-65	13	Mannheim	-5	-24
3	Braunschweig	46	79	14	München	70	-74
4	Bremen	8	111	15	Nürnberg	59	-26
5	Essen	-36	52	16	Passau	114	-56
6	Freigburg	-22	-76	17	Regensburg	83	-41
7	Hamburg	34	129	18	Saarbrücken	-40	-28
8	Hof	74	6	19	Würzburg	31	-12
9	Karlsruhe	-6	-41	20	Bielefeld	0	71
10	Kassel	21	45	21	Lübeck	50	140
11	Kiel	37	155	22	Münster	-20	70

5.6 Strengths and Limitations of the Standardized Within-Cluster Criterion

When **A** is a matrix of squared Euclidean distances, criterion (2.3) is well-recognized as the important within-cluster sum of squares criterion. Although this is certainly one of the most frequently used measures for partitioning under such conditions, there is a noteworthy caveat. The within-cluster sum of squares criterion tends to produce spherical clusters (e.g., of a circular shape in the two-dimensional context), even when clusters of this shape are not necessarily appropriate. For example, in certain biological and medical applications, clusters are long and thin (e.g., of an elliptical shape in the two-dimensional context), and, therefore, they are apt to be poorly recovered by the within-cluster sum of squares criterion.

Table 5.6. Results for German towns data at different numbers of clusters (K), criterion (2.3).

K	Standardized within-cluster sums	$ssrr(K)$	Partition
2	64409.45	--	{1,3,4,5,7,10,11,12,20,21,22} {2,6,8,9,13,14,15,16,17,18,19}
3	39399.14	38.83	{1,5,6,9,12,13,18} {2,8,14,15,16,17,19} {3,4,7,10,11,20,21,22}
4	21719.32	44.87	{1,5,10,12,20,22} {2,8,14,15,16,17,19} {3,4,7,11,21} {6,9,13,18}
5	16592.55	23.60	{1,5,10,12,20,22} {2,14,16,17} {3,4,7,11,12} {6,9,13,18} {8,15,19}
6	11889.25	28.35	{1,5,12,22} {2,14,16,17} {3,10,20} {4,7,11,21} {6,9,13,18} {8,15,19}
7	9950.75	16.30	{1,5,12} {2,14,16,17} {3,10} {4,20,22} {6,9,13,18} {7,11,21} {8,15,19}
8	8177.50	17.82	{1,5,12} {2,14} {3,10} {4,20,22} {6,9,13,18} {7,11,21} {8,15,19} {16,17}

The important consideration of the appropriateness of spherical clusters notwithstanding, (2.3) remains a widely used criterion for many applications. The branch-and-bound paradigm described herein can provide optimal solutions for hundreds of objects into as many as eight clusters, provided that the clusters are well-separated. For randomly generated data with no inherent cluster structures, the approach becomes computationally infeasible for approximately 60 objects in 6 clusters.

Criterion (2.3) can be used even when **A** does not have metric properties. In other words, there is no reason that the standardized within-cluster sums of dissimilarities could not be employed for any dissimilarity matrix. However, as described in section 5.1, Components 2 and 3 of the bound are no longer valid. This greatly diminishes the ability to obtain optimal solutions for problems of practical size. In fact, there appears to be no definitive advantage to using (2.3) as opposed to (2.2) when **A** does not have metric properties.

5.7. Available Software

We have made available a software program for minimizing the within-cluster sum of squares, *bbwcss.for*, which can be accessed and downloaded from http://www.psiheart.net/quantpsych/monograph.html or http://garnet.acns.fsu.edu/~mbrusco. The program *bbwcss.for* is similar to the algorithm for minimizing (2.3) originally described by Koontz et al. (1975) and, later, by Diehr (1985). The algorithm also incorporates a relabeling of the objects that separates nearest neighbors prior to implementation of the branch-and-bound process. This relabeling step, along with a sequential (staged) solution process, enables significant improvements in computational speed (see Brusco, 2005).

The file structure for *bbwcss.for* is identical to *bbdiam.for*, *bbdisum.for*, and *bbwcsum.for*. However, unlike these other programs, *bbwcss.for* should only be used when the input matrix contains squared Euclidean distances between pairs of objects. This caution cannot be overemphasized. If the program is applied to a generic dissimilarity matrix, the results are apt to be incorrect because the strong bounding procedures used in the branch-and-bound algorithm are no longer valid.

To illustrate the operation of *bbwcss.for*, we present the following screen display information for execution of the program for the data in Table 5.5, which contains coordinates for 22 German towns as reported by Späth (1980, p. 43). The screen display is for the four-cluster solution that was interpreted in section 5.5.

The input information is as follows:

```
> TYPE 1 FOR HALF MATRIX OR TYPE 2 FOR FULL MATRIX
> 1
> PLEASE INPUT NUMBER OF CLUSTERS 2 TO 10
> 4
```

The output information is as follows:

```
NUMBER OF OBJECTS    5 Z =      774.50000************
NUMBER OF OBJECTS    6 Z =     1403.50000    1403.50010
NUMBER OF OBJECTS    7 Z =     3107.66667    3107.66677
NUMBER OF OBJECTS    8 Z =     3827.66667    3827.66677
NUMBER OF OBJECTS    9 Z =     5767.00000    5767.00010
NUMBER OF OBJECTS   10 Z =     5964.00000    5964.00010
NUMBER OF OBJECTS   11 Z =     6903.00000    6903.00010
NUMBER OF OBJECTS   12 Z =     8762.33333    8762.33343
NUMBER OF OBJECTS   13 Z =    10490.66667   10490.66677
```

74   5 Minimum Within-Cluster Sums of Squares Partitioning

```
NUMBER OF OBJECTS   14 Z =    12957.33333 12957.33343
NUMBER OF OBJECTS   15 Z =    16102.83333 16102.83343
NUMBER OF OBJECTS   16 Z =    18637.00000 18637.00010
NUMBER OF OBJECTS   17 Z =    19108.50000 19374.30010
NUMBER OF OBJECTS   18 Z =    20112.10000 20112.10010
NUMBER OF OBJECTS   19 Z =    20161.40000 20161.40010
NUMBER OF OBJECTS   20 Z =    21266.59048 21266.59058
NUMBER OF OBJECTS   21 Z =    21517.15714 21517.15724
NUMBER OF OBJECTS   22 Z =    21719.32381 21719.32391

MINIMUM WITHIN CLUSTER SUM OF SQUARES    21719.32381
TOTAL CPU TIME                   0.00
  2   3   4   4   2   1   4   3   1   2   4   2   1   3   3   3   3
  1   3   2   4   2
Stop - Program terminated.
```

Like *bbwcsum.for* in Chapter 4, *bbwcss.for* produces summary output that enables tracking of the solution process. For each stage of the process, the current number of objects, optimal within-cluster sums of squares, and upper bound are provided. For this particular data set, tracking solution progress was not necessary, as the required CPU time of "0.00" indicates that solution time for the whole process was less than .01 second. The cluster assignments for each town are consistent with those reported in Table 5.6.

With regard to scalability for the same squared Euclidean distance matrix, the efficiency of *bbwcss.for* falls somewhere between *bbdiam.for* and *bbwcsum.for*. Although nowhere near as efficient as *bbdiam.for*, *bbwcss.for* is typically much faster than *bbwcsum.for*, and can produce guaranteed optimal solutions for rather sizable values of n and K. To illustrate, consider the following five-cluster solution for a data set comprised of coordinates for $n = 59$ German towns, as reported by Späth (1980, p. 80).

The input information is as follows:

```
> TYPE 1 FOR HALF MATRIX OR TYPE 2 FOR FULL MATRIX
> 1
> PLEASE INPUT NUMBER OF CLUSTERS 2 TO 10
> 5
```

The output information is as follows:

```
NUMBER OF OBJECTS    6 Z =       65.00000************
```

5.7. Available Software

```
NUMBER OF OBJECTS    7 Z =      87.50000    87.50010
NUMBER OF OBJECTS    8 Z =     251.75000  3741.16677
NUMBER OF OBJECTS    9 Z =     672.25000   876.75010
NUMBER OF OBJECTS   10 Z =     875.08333   875.08343
NUMBER OF OBJECTS   11 Z =    1500.08333  2188.08343
NUMBER OF OBJECTS   12 Z =    1587.08333  1587.08343
NUMBER OF OBJECTS   13 Z =    2783.58333  2783.58343
NUMBER OF OBJECTS   14 Z =    3293.83333  3293.83343
NUMBER OF OBJECTS   15 Z =    3368.00000  3368.00010
NUMBER OF OBJECTS   16 Z =    3928.16667  3928.16677
NUMBER OF OBJECTS   17 Z =    5635.66667  5698.00010
NUMBER OF OBJECTS   18 Z =    6134.60714  6358.16677
NUMBER OF OBJECTS   19 Z =    7142.70714  7142.70724
NUMBER OF OBJECTS   20 Z =    7426.43333  7496.15724
NUMBER OF OBJECTS   21 Z =    9337.07500  9951.50010
NUMBER OF OBJECTS   22 Z =   10420.33333 10510.54177
NUMBER OF OBJECTS   23 Z =   13041.16667 13041.16677
NUMBER OF OBJECTS   24 Z =   13443.64286 13443.64296
NUMBER OF OBJECTS   25 Z =   14659.22619 14694.14296
NUMBER OF OBJECTS   26 Z =   15167.47619 15318.47629
NUMBER OF OBJECTS   27 Z =   16199.00000 16203.47629
NUMBER OF OBJECTS   28 Z =   16703.40000 16703.40010
NUMBER OF OBJECTS   29 Z =   18590.23333 18590.23343
NUMBER OF OBJECTS   30 Z =   18930.46548 18930.46558
NUMBER OF OBJECTS   31 Z =   19919.10833 19919.10843
NUMBER OF OBJECTS   32 Z =   20260.85278 20546.37510
NUMBER OF OBJECTS   33 Z =   20954.24167 21106.65288
NUMBER OF OBJECTS   34 Z =   22283.76548 22283.76558
NUMBER OF OBJECTS   35 Z =   25516.64643 25516.64653
NUMBER OF OBJECTS   36 Z =   26119.89048 26718.16042
NUMBER OF OBJECTS   37 Z =   27142.98701 27158.35724
NUMBER OF OBJECTS   38 Z =   28335.88889 29119.27283
NUMBER OF OBJECTS   39 Z =   28703.98889 28703.98899
NUMBER OF OBJECTS   40 Z =   29324.22698 29324.22708
NUMBER OF OBJECTS   41 Z =   30195.58153 30195.58163
NUMBER OF OBJECTS   42 Z =   30603.14416 30664.88520
NUMBER OF OBJECTS   43 Z =   31844.06557 32066.45486
NUMBER OF OBJECTS   44 Z =   31928.46667 31928.46677
NUMBER OF OBJECTS   45 Z =   31963.19881 31963.19891
NUMBER OF OBJECTS   46 Z =   32294.07976 32294.07986
NUMBER OF OBJECTS   47 Z =   33660.98452 33660.98462
NUMBER OF OBJECTS   48 Z =   34627.86634 34627.86644
NUMBER OF OBJECTS   49 Z =   35246.68452 35246.68462
NUMBER OF OBJECTS   50 Z =   35387.25397 35387.25407
NUMBER OF OBJECTS   51 Z =   36414.46230 36414.46240
NUMBER OF OBJECTS   52 Z =   36708.76587 36708.76597
```

```
NUMBER OF OBJECTS   53 Z =   36994.42028  36994.42038
NUMBER OF OBJECTS   54 Z =   37110.27976  37110.27986
NUMBER OF OBJECTS   55 Z =   37739.12592  37739.12602
NUMBER OF OBJECTS   56 Z =   37904.68147  37904.68157
NUMBER OF OBJECTS   57 Z =   38061.18732  38061.18742
NUMBER OF OBJECTS   58 Z =   38323.91459  38323.91469
NUMBER OF OBJECTS   59 Z =   38716.01986  38716.01996
MINIMUM WITHIN CLUSTER SUM OF SQUARES   38716.01986
TOTAL CPU TIME                          9.99
  5  1  1  4  3  1  3  1  4  4  3  3  3  1  5  1
  1  5  2  3  5  1  1  2  4  5  4  3  5  2  1  2
  3  2  2  1  4  4  2  1  1  1  2  3  2  2  1  1
  1  2  5  4  3  1  3  1  3  3  1
Stop - Program terminated.
```

Using (2.1), the number of feasible partitions of $n = 59$ cities into $K = 5$ clusters is more than 1.44×10^{39}. Nevertheless, an optimal partition is obtained in approximately 10 seconds of microcomputer CPU time. Brusco (2005) reported optimal solutions for up to $K = 8$ clusters for these same data. Perhaps even more notable is the fact that optimal partitions for Fisher's iris data (Fisher, 1936), which consists of $n = 150$ objects (plants), were obtained for $2 \leq K \leq 6$ clusters.

6 Multiobjective Partitioning

6.1 Multiobjective Problems in Cluster Analysis

A number of authors have addressed, in various ways, issues pertaining to multiobjective partitioning (Brusco, Cradit, & Stahl, 2002; Brusco, Cradit, & Tashchian, 2003; Delattre & Hansen, 1980; DeSarbo & Grisaffe, 1998; Ferligoj & Batagelj, 1992; Guénoche, 2003; Krieger & Green, 1996). There are a variety of possible circumstances under which the construction of a partition of objects based on two or more criteria might be especially appropriate. Broadly defined, these circumstances can be divided into two distinct categories: (a) partitioning of objects using multiple bases, and (b) partitioning of objects in accordance with a single base but using multiple criterion. In the subsections that follow, we present an example for each of these two categories.

6.2 Partitioning of an Object Set Using Multiple Bases

Partitioning of objects using multiple bases occurs when we consider multiple sets of data in the clustering of the objects and we cannot legitimately lump the multiple sets into a single data set and apply traditional clustering procedures. An excellent illustration of this situation within the context of market segmentation is described by Krieger and Green (1996). To consider an analogous example, suppose that a long distance service provider (company A) has access to a survey of potential business customers regarding the importance of various factors related to the provision of long distance service. Information on firm demographics (firmographics), such as assets, number of employees, return on investment, sales, etc. might also be available. Company A also has data pertaining to the satisfaction of the firms with respect to their current provider, as well as the propensity to switch.

Company A could aggregate all of the variable measurements into a single data set and apply standard partitioning procedures to obtain clusters of the firms. What is more appropriate, however, is to develop clusters that are both homogeneous with respect to the firmographic measures, yet also explain variation in current levels of satisfaction or switching. Research has shown that such competing objectives are often somewhat adversarial (Brusco et al., 2002, 2003; DeSarbo & Grisaffe, 1998; Krieger & Green, 1996). In other words, in the context of our example, partition optimization procedures based on firmographic homogeneity yield homogeneous clusters, but very poor explanation of satisfaction or switching potential. On the other hand, clustering based on explaining variation in satisfaction or switching generally produces clusters that are not homogeneous with respect to the firmographic measures. This lack of homogeneity increases the difficulty of preparing targeted marketing efforts at selected groups of firms. What is necessary is a method that can simultaneously provide clusters that are homogeneous with respect to firmographic measures, yet also explain exogenous factors such as satisfaction or switching behavior. Multiobjective programming can help to achieve this goal.

To illustrate the concept, we use the context of the example from Brusco et al. (2003), however, the data set for illustration purposes was synthetically constructed. Suppose that company A has Likert-scale measurements on four measures (service attributes) that are to be used as clustering variables: billing accuracy (v_1), responsive account executives (v_2), low-cost services (v_3), and innovative products (v_4). Company A would like to identify clusters that are homogeneous with respect to these measures (in other words, group similar potential customers with respect to their attitudes toward service attributes), yet also explain variation for two Likert-scale performance measures related to satisfaction: overall satisfaction with current service provider (w_1) and willingness to switch service providers (w_2). Variable measurements for 20 firms are presented in Table 6.1.

Two separate 20 × 20 dissimilarity matrices, $\mathbf{A}v$ and $\mathbf{A}w$, were developed by computing the squared Euclidean distances between pairs of firms using the service attribute and performance measure variables, respectively. For example, matrix entry a_{12} of $\mathbf{A}v$ is $(3 - 6)^2 + (7 - 6)^2 + (4 - 5)^2 + (7 - 6)^2 = 12$ and matrix entry a_{12} of $\mathbf{A}w$ is $(4 - 6)^2 + (6 - 3)^2 = 13$. Because the dissimilarity matrices correspond to squared Euclidean distances, criterion (2.3) was selected as the appropriate partitioning index. To obtain baselines, we begin by constructing optimal partitions (under the assumption of $K = 4$ clusters) using the service attributes (v_1, v_2, v_3, v_4) and the performance measures (w_1, w_2), independently.

Table 6.1. Example data set (20 objects measured on four performance drivers and two performance measures).

Object	v_1	v_2	v_3	v_4	w_1	w_2
1	3	7	4	7	4	6
2	6	6	5	6	6	3
3	4	7	7	7	7	2
4	1	7	3	6	3	5
5	5	6	6	7	1	7
6	7	3	7	4	2	4
7	6	2	7	2	6	1
8	7	6	7	4	7	3
9	7	4	6	1	7	2
10	6	7	7	5	5	3
11	5	7	6	2	2	7
12	2	7	7	5	7	3
13	3	6	7	3	7	1
14	7	6	7	2	3	5
15	1	7	6	4	5	3
16	7	2	5	6	6	2
17	6	1	3	7	7	2
18	5	5	2	7	4	6
19	7	3	6	5	7	4
20	6	4	4	7	2	4

The optimal partitioning solutions for these two sets of variables are shown in the first two panels of Table 6.2. Our example has $L = 2$ data sources: (a) service attributes, and (b) performance measures. Therefore, we can conveniently express the within-cluster sums of squares as a percentage of the total sum of squares, which is computed as follows:

$$TSS_l = \frac{\sum_{i<j} a_{ij}^l}{n} \quad \text{for } 1 \leq l \leq L, \tag{6.1}$$

where a_{ij}^l is the squared Euclidean distance between objects i and j for data source l.

The optimal within-cluster sums of squares for the service attribute data is 96.917, which corresponds to explanation of 66.2% of the total variation in these data (see equation 6.7). Unfortunately, this partition explains only 4.4% of the total variation for the performance measures. Similarly, the optimal partition for the performance measures explains

88.7% of the total variation in these data, but this same partition explains only 13.8% of the variation for service attributes. Thus, the two sets of clustering variables appear to have a somewhat antagonistic relationship. If we select an optimal partition for the service attributes, the explanation of variation in the performance measures is extremely poor, and vice versa. The question is: Can a multiobjective approach identify a partition that yields a good compromise between the two sets of data, yielding good explanatory power for both sets of variables?

Table 6.2. Single-objective and biobjective solution for the data in Table 6.1.

Criterion optimized	$F(\lambda^*)$	Explained variation (v_1, v_2, v_3, v_4)	Explained variation (w_1, w_2)	Partition (λ^*)
Service attributes	N / A	66.2%	4.4%	{1,2,3,5,10} {4,12,13,15} {6,7,8,9,11,14} {16,17,18,19,20}
Performance measures	N / A	13.8%	88.7%	{1,4,6,14,18,20} {2,8,10,12,15,19} {3,7,9,13,16,17} {5,11}
Biobjective equal weights	84.54286	51.4%	78.1%	{1,4,5,18,20} {2,3,8,10,12,13,15} {7,9,16,17,19} {6,11,14}

When there are $l = 1,..., L$ separate sources of data, the general multiobjective model considered for this type of problem is as follows:

$$\text{Minimize}: Z_1(\pi_K) = \sum_{l=1}^{L} w_l f_3^{l}(\pi_K), \quad \pi_K \in \Pi_K \quad (6.2)$$

$$s.t. \quad f_3^{l}(\pi_K) \le Q_3^{l} \quad \text{for } 1 \le l \le L; \quad (6.3)$$

$$\sum_{l=1}^{L} w_l = 1; \quad (6.4)$$

$$w_l > 0 \quad \text{for } 1 \le l \le L. \quad (6.5)$$

The objective function of the model is a weighted function of within-cluster sums of squares indices for the L data sources. Constraints (6.3) are optional, but can be used to place upper restrictions, Q_3^l, on the within-cluster sums of squares for the data sources. Constraints (6.4) and (6.5) require a convex combination of nonzero weights that sum to 1. This type of multiobjective model draws heavily from the work of Soland (1979), and is consistent with previous applications of multiobjective partitioning in the marketing literature (Brusco et al., 2002, 2003; DeSarbo & Grisaffe, 1998). There are many possible variations on the multiobjective model. For example, the objective function can be normalized using the optimal within-cluster sums of squares for each of the data sources. Specifically, if f_l^* is the optimal value of $f_3(\pi_K)$ for data source l ($1 \leq l \leq L$), then an alternative objective function is:

$$\text{Minimize}: Z_2(\pi_K) = \sum_{l=1}^{L} \left(\frac{w_l}{f_l^*}\right)\left(f_3^l(\pi_K)\right). \quad (6.6)$$
$$\pi_K \in \Pi_K$$

Objective function (6.6) is bounded below by 1, which would occur when all data sources achieved their optimal index values. In most cases, however, we will have $f_3^l(\pi_K) > f_l^*$ for one or more l. Yet another possibility is to maximize a weighted function of the proportion of explained variation for the data sources. The proportion of explained variation for data source l is:

$$\Phi_l(\pi_K) = \frac{f_3^l(\pi_K)}{TSS_l} \quad \text{for } 1 \leq l \leq L. \quad (6.7)$$

The resulting objective function would then be denoted as:

$$\text{Maximize}: Z_3(\pi_K) = \sum_{l=1}^{L} w_l \Phi_l(\pi_K). \quad (6.8)$$
$$\pi_K \in \Pi_K$$

One of the advantages of a multiobjective model based on (6.8) is that the objective function will be bounded between 0 and 1. This condition holds because the constraints (6.4) and (6.5) require a convex combination of weights and the fact that the Φ_l values from (6.7) are between 0 and 1.

Because our particular example has two data sources, it is a biobjective partitioning problem. The branch-and-bound algorithm for (2.3) requires only minor modifications for the biobjective problem. Specifically, the partial solution evaluation, branching, and retraction steps must appropriately update within-cluster sums for two matrices (in this case **Av** and **Aw**) instead of one. The within-cluster sums of squares for both matrices are appropriately modified by user-specified weights. Any solution produced by the algorithm is said to be *nondominated*. A partition, π_K, is nondominated if no other feasible K-cluster partition exists with criteria values as good or better than those of π_K, with at least one criterion that is strictly better.

The third panel of Table 6.2 presents a biobjective partition for equal weights ($w_1 = w_2 = .5$). This partition provides an exceptionally good trade-off between the individual partitions for service attributes and performance measures. As noted previously, the optimal partition for performance measures explained 88.7% of the total variation for those two variables; however, the partition explained only 13.8% of the variation in the service attributes data. The biobjective partition requires a modest sacrifice in explained variation for the performance measures (88.7% to 78.1%) for a tremendous improvement in explained variation for the service attributes (13.8% to 51.4%). We can also view the compromise from the position of the service attributes. There is a modest reduction in explanation of variation for service attributes (66.2% to 51.4%), which is sacrificed for a marked improvement in explained variation for performance measures (4.4% to 78.1%). In short, the biobjective approach has yielded an excellent compromise solution for the data sources, despite the somewhat antagonistic indications displayed by their individual optimal partitions.

6.3 Partitioning of Objects in a Single Data Set Using Multiple Criteria

In Chapters 3 and 4, we observed that the three-cluster partition for the lipread consonants data based on (2.1) was comparable, but somewhat different from the partition of the same data set based on (2.2). In other situations, two different criteria can produce significantly different partitions, and an analyst might have difficulty ascertaining which criterion (and corresponding partition) is more appropriate. To facilitate this decision, the analyst might seek a partition that enables a trade-off among the two criteria in question.

6.3 Partitioning of Objects in a Single Data Set Using Multiple Criteria

Multiobjective partitioning can also be particularly beneficial when using criteria that are prone to result in myriad optimal solutions. Such is the case for the partition diameter index (2.1). To select from among the set of alternative optima for the diameter criterion, a biobjective approach can be employed. In the first stage, an optimal diameter is identified using the branch-and-bound algorithm. In the second stage, a second criterion (such as 2.2) is optimized subject to the constraint that the partition diameter criterion does not worsen by more than some user-prescribed limit. To illustrate, we return to the lipread consonants example. The optimal partition diameter for $K = 3$ clusters is $f_1^*(\pi_3) = 306$, and the corresponding within-cluster sum of dissimilarities for the solution in Table 3.1 is $f_2(\pi_3) = 15590$. On the other hand, the optimal within-cluster sums of pairwise dissimilarities from Table 4.1 is $f_2^*(\pi_3) = 13177$, but the corresponding diameter of the partition associated with this optimum is $f_1(\pi_3) = 363$.

A biobjective program was applied to optimize (2.2), subject to a constraint for (2.1). This model is a slightly different approach to a bicriterion problem from the models previously described. Because we use a required value for the first criterion as a constraint, there is no issue of weights. The algorithm was applied for various partition diameters, and the results are displayed in Table 6.3. The first row of Table 6.3 was obtained by minimizing (2.2) subject to a constraint that partition diameter achieves the optimal three-cluster value of $f_1^*(\pi_3) = 306$. The partition is the same as the one reported in Table 3.3, and thus there are no minimum-diameter partitions that achieve values for $f_2(\pi_3)$ that are better than 15590. Perhaps even more interesting is the fact that this solution remains optimal for diameter constraints up to $f_1(\pi_3) = 324$. When the diameter restriction is relaxed to 325, some improvement in (2.2) is realized by moving k from cluster 2 to cluster 3. Nevertheless, to obtain any improvement in the within-cluster sums of dissimilarities, a big penalty in partition diameter is necessarily incurred.

A steady improvement in (2.2) obtained for sacrifices in (2.1) is observed for remaining rows of Table 6.3. Cluster 1 remains quite stable throughout the partitions in Table 6.3. In fact, there is no change in cluster 1 across the last seven rows of the table. The changes occurring in clusters 2 and 3 are particularly interesting. As there is gradual easing of the restriction on partition diameter, the number of objects assigned to clusters 2 and 3 is more balanced, which is expected for criterion (2.2). In some cases, changes from one row to the next are rather small. For example, row 3 is related to row 2 by the simple movement of v from cluster 1 to cluster 3.

In other situations significant changes in the partitions produce rather small changes in the two objective criteria. For example, row 7 is obtained by constraining partition diameter to no more than 347, whereas the constraint for row 8 is 349. The result is that the two unit increase in partition diameter yields only a small improvement in the total within-cluster sums of pairwise dissimilarities, from 13706 to 13428. Although these criterion changes are quite modest, the partition changes that produced them are fairly substantial, consisting of the movement of four objects (objects j and k are moved from cluster 2 to cluster 3, and objects h and n are moved from cluster 3 to cluster 2.

Table 6.3. Biobjective solutions for the lipread consonants data.

Row	Diameter (2.1)	Sums (2.2)	Partition
1	306	15590	{b,c,d,g,j,p,t,v,z} {f,h,k,*l*,m,n,r,s,x} {q,w,y}
2	325	14569	{b,c,d,g,j,p,t,v,z} {f,h,*l*,m,n,r,s,x} {k,q,w,y}
3	329	14229	{b,c,d,g,j,p,t,z} {f,h,*l*,m,n,r,s,x} {k,q,v,w,y}
4	335	13944	{b,c,d,g,p,t,v,z} {f,h,j,*l*,n,r,s,x} {k,m,q,w,y}
5	337	13932	{b,c,d,g,p,t,v,z} {f,h,j,*l*,n,s,y} {k,m,q,r,w,x}
6	345	13778	{b,c,d,g,p,t,v,z} {f,j,*l*,n,r,s,x} {h,k,m,q,w,y}
7	347	13706	{b,c,d,g,p,t,v,z} {f,j,k,*l*,r,s,x} {h,m,n,q,w,y}
8	349	13428	{b,c,d,g,p,t,v,z} {f,h,*l*,n,r,s,x} {j,k,m,q,w,y}
9	355	13382	{b,c,d,g,p,t,v,z} {h,k,*l*,n,r,s,x} {f,j,m,q,w,y}
10	363	13177	{b,c,d,g,p,t,v,z} {f,h,*l*,m,n,s,x} {j,k,q,r,w,y}

6.4 Strengths and Limitations

The multiobjective programming paradigm permits significant flexibility in cluster analyzing data. For a single data set, considerable confidence can be gained by finding a partition that produces excellent index values on two or more criteria. When multiple bases for cluster analysis are available, multiobjective programming can help to ensure that each of the bases is awarded sufficient consideration in the clustering process. In the absence of multiobjective programming, one or more of the bases would likely tend to dominate the solution if the multiple bases were collapsed into a single data set for analysis purposes. A multiobjective perspective does not necessarily increase the difficulty of obtaining optimal solutions for partitioning problems. In fact, in some instances, the solution time is considerably improved. For example, the demonstration in section 6.3 revealed that using the diameter measure as the primary criterion led to considerable improvement in the solution time for (2.2) when the diameter constraint was imposed.

The principal limitation of multiobjective programming is the selection of an appropriate weighting scheme. The severity of this problem increases markedly when three or more criteria are considered. For biobjective scenarios, weighting the first criterion from 0 to 1 in increments of .05 or .1, and setting the corresponding weight for criterion 2 as 1 minus the weight for criterion 1, is a particularly effective approach. Unfortunately, this type of parameter search becomes computationally infeasible as the number of criteria increases.

6.5 Available Software

We have made available a software program for biobjective partitioning based on the within-cluster sums of squares criterion, *bbbiwcss.for*. Like all other programs for this monograph, this can be accessed and downloaded from both http://www.psiheart.net/quantpsych/monograph.html or http://garnet.acns.fsu.edu/~mbrusco. This program reads two separate $n \times n$ distance matrices from the files *amat.dat* and *bmat.dat*, respectively. The user is prompted as to whether the format of the data in both files is a half matrix or a full matrix (the program currently requires both matrices to be in the same format, but this could easily be modified). The user is also prompted for the number of clusters, as well as weights for the within-cluster sums of squares index for each matrix. We recommend that the weights be specified as nonnegative values that sum to 1, in accordance with practice in multiobjective clustering applications (Brusco et al., 2002, 2003; DeSarbo & Grisaffe, 1998). However, the program will run for any pair of weights.

To illustrate the operation of *bbbiwcss.for*, we present the following screen display information for execution of the program for the data in Table 6.1. The screen display is for the four-cluster solution that was interpreted in section 6.2. Also, observe that selected weights for the displayed solution are ($w_1 = w_2 = .5$), yielding an equal weighting scheme.

The input information is as follows:

```
> TYPE 1 FOR HALF MATRIX OR 2 FOR FULL MATRIX INPUT
> 1
> INPUT NUMBER OF CLUSTERS
> 4
> INPUT WEIGHTS
> .5 .5
```

The output information is as follows:

```
NUMBER OF OBJECTS    5 Z =         2.00000************
NUMBER OF OBJECTS    6 Z =         5.50000        15.75010
NUMBER OF OBJECTS    7 Z =         9.83333        14.50010
NUMBER OF OBJECTS    8 Z =        13.58333        13.58343
NUMBER OF OBJECTS    9 Z =        15.16667        15.16677
NUMBER OF OBJECTS   10 Z =        17.91667        17.91677
NUMBER OF OBJECTS   11 Z =        25.20833        25.20843
NUMBER OF OBJECTS   12 Z =        37.00000        37.00010
NUMBER OF OBJECTS   13 Z =        43.58333        43.87510
NUMBER OF OBJECTS   14 Z =        49.95833        50.43343
NUMBER OF OBJECTS   15 Z =        57.20833        57.20843
NUMBER OF OBJECTS   16 Z =        63.70833        63.70843
NUMBER OF OBJECTS   17 Z =        73.33333        73.33343
NUMBER OF OBJECTS   18 Z =        77.30833        77.30843
NUMBER OF OBJECTS   19 Z =        81.96786        82.95843
NUMBER OF OBJECTS   20 Z =        84.54286        84.54296

WCSS MATRIX A =           136.60952
WCSS MATRIX B =            32.47619
TOTAL CPU TIME                 5.96
 1  2  2  1  1  3  4  2  4  2  3  2  2  3  2  4  4
 1  4  1
Stop - Program terminated.
```

The solution output is similar to *bcwcsum.for* and *bbwcss.for* for Chapters 4 and 5, respectively. However, the values reported at each stage are the optimal weighted within-cluster sums of squares values and the bound for that value. The within-cluster sums of squares values that correspond to the optimal biobjective solution are also reported. In other words, the within-cluster sums of squares values for matrices A and B are 136.60952 and 32.47619, respectively, and the optimal weighted objective function value is .5 (136.60952) + .5 (32.47619) = 84.54286. The total CPU time required to obtain the optimal solution to the biobjective problem was 5.96 seconds, and the optimal partition shown is consistent with the solution in Table 6.2 corresponding to equal weights.

There are numerous possible modifications of *bbbiwcss.for*. First, the code could be modified to run the biobjective algorithm for a systematic set of weights rather than require user specification of the weights. For example, an outer loop represented by *iloop* = 0 to 10 could embed the algorithmic code. The weight for matrix A would be *iloop* / 10 and the weight for matrix B would be (1 – *iloop* / 10). Thus, in total, 11 solutions would be produced. Two of these are the single objective optimal solutions for each matrix and the others differentially weighting the matrices. A second modification, as described in section 6.2, would be to normalize the within-cluster sums of squares for each matrix based on the optimal single criterion within-cluster sums of squares. A third extension of the code would be to incorporate three or more matrices. Although this is not particularly difficult from a computational standpoint, finding a good set of weights when three or more criteria are considered can be a time-consuming process.

Part II

Seriation

7 Introduction to the Branch-and-Bound Paradigm for Seriation

7.1 Background

Although seriation is a natural tendency in many mundane tasks such as listing names in alphabetical order or prioritizing daily chores in "order of importance," the scientific applications are usually more demanding. Archaeological finds may need to be put into chronological order (Kendall, 1971a, 1971b; Robinson, 1951), psychological data of sensory recognition of stimuli (Cho, Yang, & Hallett, 2000; Morgan, Chambers, & Morton, 1973; Vega-Bermudez, Johnson, & Hsiao, 1991) can be ordered to examine flow of confusion, or astronomical bodies may need to be ordered by their distance from an observation point (usually Earth). Seriation methods allow analysts to find structure in otherwise obscure or disorderly proximity matrices.

The goal of seriation methods is to find an optimal permutation to more clearly reveal structure among the stimuli. Seriation methods permute rows and columns simultaneously to find a good structural representation to proximity matrices as measured by an appropriate index. An objective function is formulated according to a criterion specified by the analyst. Mathematical models are designed to maximize or minimize the objective function. Using appropriate criteria for the data at hand, seriation methodologies are designed to order the objects to optimize the patterning within the data and describe how well the data corresponds to patterning ideals. As described in Chapters 8 and 9, an analyst can maximize the dominance index or inquire as to the degree of adherence to gradient flows within rows and/or columns, such as resemblance to anti-Robinson structure. By using indices to determine the degree of adherence to gradient flows within rows and/or columns, we uncover a vital matrix structure. More than ordering the objects, seriation can relate the objects to one another in terms of "distance" to produce a graphical representation of the relationships between the objects.

Like the traveling salesman problem and workstation sequencing problem described in Chapter 1, seriation of proximity data presents a challenging combinatorial optimization problem. In light of its elegant mathematical aspects and its relationship to other important sequencing and linear ordering problems, seriation and related topics have received a significant amount of attention in the research literature (Barthélemy, Hudry, Isaak, Roberts, & Tesman, 1995; Brusco, 2001, 2002a, 2002b, 2002c; Brusco & Stahl, 2001a, 2004; Charon, Guénoche, Hudry, & Woirgard, 1997; Charon, Hudry, & Woirgard, 1996; DeCani, 1969, 1972; Flueck & Korsh, 1974; Grötschel, Jünger, & Reinelt, 1984; Hubert, 1974, 1976; Hubert & Golledge, 1981; Hubert & Arabie, 1986; Hubert et al., 2001).

Factors in choosing whether to use an *optimal* or *heuristic* procedure to uncover structure in proximity matrices include (1) the number of objects, (2) the selected seriation criterion, and (3) the structural properties of the data set. Optimal procedures have gained in popularity over the years with the advent of the computer age, facilitating file-sharing between researchers and greatly increasing computer memory storage on personal computers, in conjunction with ever-improving methodological techniques. Several procedures exist for finding optimal solutions to seriation problems.

One of the most effective optimal solution procedures for seriation is dynamic programming, which was originally suggested by Lawler (1964), drawing from the work of Bellman (1962) and Held and Karp, (1962), within the context of minimum feedback arc sets in engineering. Dynamic programming has been adapted for seriation in data analysis problems (Brusco, 2002c; Brusco & Stahl, 2001a; Hubert & Arabie, 1986; Hubert & Golledge 1981; Hubert et al., 2001). Hubert et al. (2001) provide a comprehensive review of dynamic programming applications in seriation and have also made available a number of Fortran programs for dynamic programming implementation. Although dynamic programming is robust with respect to the properties of the proximity data and can be adapted for a variety of seriation criteria, the approach is constrained by computer memory storage limitations. For example, on a microcomputer with 1 GB of RAM, dynamic programming would be limited to problems with roughly 26 or fewer objects.

Some seriation problems can be effectively modeled using integer programming methods (Brusco, 2001, 2002a; DeCani, 1972) or cutting plane methods (Grötschel et al., 1984). Unlike dynamic programming, however, these methods tend to be more sensitive to the properties of the proximity data. Furthermore, such methods often require extensive familiarity with commercial mathematical programming packages and/or considerable programming acumen.

Branch-and-bound methods are also sensitive to the properties of the proximity data but have proven to be an effective alternative to dynamic programming for a variety of seriation problems (Brusco, 2002b; Brusco & Stahl, 2004, DeCani, 1972; Flueck & Korsh, 1974). The principal advantage of branch-and-bound relative to dynamic programming is that the former requires minimal storage and can thus be applied to proximity matrices that are too large for dynamic programming implementation. For well-structured proximity matrices, branch-and-bound is often an efficient and scalable approach, capable of handling proximity matrices with 30, 40, or more objects in some instances. Poorly structured data can be easily generated to induce very poor performance by a branch-and-bound method. However, most analysts are not interested in seriation of this type of data.

To summarize, dynamic programming methods for seriation are more robust than their branch-and-bound counterparts with respect to the structural properties of the proximity matrix. Branch-and-bound can be more efficient that dynamic programming for well-structured proximity matrices and can be applied to larger matrices than dynamic programming because of its modest storage requirements. Nevertheless, dynamic programming, branch-and-bound, cutting planes, integer programming, and other optimal solution procedures are ultimately limited by problem size, and heuristics are required for large problems.

7.2 A General Branch-and-Bound Paradigm for Seriation

The general, forward-branching, branch-and-bound algorithm for seriation that searches for optimal permutations can be described in ten steps with three essential components—permutation generation, evaluation, and fathoming. The permutation generation begins with an initialization of the permutation, the position pointer, and the initial lower bound. The initial lower bound is often found with the aid of heuristic techniques and sometimes set to -∞. (In cases of maximizing an objective function for nonnegative data, the initial lower bound can be set to 0.) However, heuristics and other procedures might use the evaluation of a particular permutation to set the initial lower bound. Therefore, the general algorithm sets the lower bound prior to initializing the permutation array. Tests or *checks* are performed to ensure that a valid permutation is being generated. Two such checks are the redundancy check to ensure that no object is chosen more than once in a permutation and the retraction test to ensure that only valid objects are chosen, i.e. if n objects define a matrix, then $n+1$ objects cannot be candidates for a position. Once a partial permutation is known to be valid, we can use fathoming tests to determine whether or not we wish to pursue building the permutation. A fathoming test, such as an adjacency test or bound test, is used to determine whether or not a partial sequence can lead to an optimal permutation. For example, the stored lower bound is used in the bound test to determine whether or not the current partial sequence could possibly lead to a better solution. If any of the fathoming tests fail, then we abandon the current partial permutation by selecting the next available valid object for the current position; otherwise, we move to the next position in the sequence and find the first available valid object for that position. By abandoning futile partial sequences, we avoid evaluating every possible permutation of objects to find the optimal solution. Finally, when an entire permutation is found as a true candidate for being an optimal solution, we evaluate the objective function for the given permutation. If we have not previously found a better solution, then we store the evaluated permutation and update our lower bound as the new best-found solution.

Throughout the chapters pertaining to seriation, the algorithm and all examples assume a maximization objective. We use the following notation to assist in discussions of permutations of objects (additional notation will be defined as needed):

n = the number of objects, indexed $i = 1,..., n$;
A = an $n \times n$ matrix containing data for the n objects, $\{a_{ij}\}$;
Ψ = the set of all $n!$ feasible permutations of the n objects;
$\psi(k)$ = the object in position k of permutation ψ, $(\psi \in \Psi)$;
p = a pointer for the object position;
f_L = lower bound used in bounding tests.

The algorithm presented below and the algorithmic pseudocode in Appendix B are designed for *forward* branching. That is, the permutations are generated from left to right. Another branching method is alternating branching in which objects are assigned to the ends of the permutation and progressively fill positions with objects until the middle positions are filled. In alternating branching, steps 2 and 3 (ADVANCE and BRANCH) are slightly modified as well as the fathoming routines, but the remainder of the algorithm—initialization and evaluation—are intact.

Step 0. INITIALIZE. Set $p = 1$, $\psi(p) = 1$, $\psi(k) = 0$ for $k = 2\ldots n$. Define f_L as a lower bound on the objective function.
Step 1. FORWARD BRANCH. Set $p = p + 1$.
Step 2. RIGHT BRANCH. Set $\psi(p) = \psi(p) + 1$.
Step 3. REDUNDANCY. If $\psi(p) = \psi(k)$ for any $k = 1,\ldots p - 1$, then go to Step 2.
Step 4. TERMINATION. If $p = 1$ and $\psi(p) > n$, then return best solution and Stop.
Step 5. RETRACTION TEST. If $p > 1$ and $\psi(p) > n$, then go to Step 10.
Step 6. COMPLETE SOLUTION TEST. If $p = n - 1$, then find the remaining unassigned object to complete ψ and go to Step 7. Otherwise, go to Step 8.
Step 7. EVALUATE. Compute $f(\psi)$. If $f(\psi) > f_L$, then set $f_L = f(\psi)$ and $\psi_B = \psi$. Go to Step 2.
Step 8. ADJACENCY TEST. If passed, go on to Step 9; otherwise, go to Step 2.
Step 9. BOUND TEST. If passed, go to Step 1; otherwise, go to Step 2.
Step 10. DEPTH RETRACTION. Set $\psi(p) = 0$, $p = p - 1$. Go to Step 2.

8 Seriation—Maximization of a Dominance Index

8.1 Introduction to the Dominance Index

The dominance index is perhaps the most widely used index for seriation of asymmetric matrices (Hubert et al., 2001) with a rich history in biometric, psychometric, and other literature bases (Blin & Whinston, 1974; Bowman & Colantoni, 1973, 1974; Brusco, 2001; Brusco & Stahl, 2001a; DeCani, 1969, 1972; Flueck & Korsh, 1974; Hubert, 1976; Hubert & Golledge, 1981; Phillips, 1967, 1969; Ranyard, 1976; Rodgers & Thompson, 1992; Slater, 1961). Essentially, maximization of the dominance index is achieved by finding a permutation that maximizes the sum of matrix elements above the main diagonal. For any pair of objects corresponding to the rows and columns of an asymmetric matrix **A**, the tendency will be to place object i to the left of object j in the sequence if $a_{ij} > a_{ji}$. Notice that the asymmetric matrix, **A**, need not be restricted to nonnegative entries. The problem of maximizing the dominance index for an asymmetric matrix **A** can be mathematically stated as follows (Lawler, 1964):

$$\max_{\psi \in \Psi} : f(\psi) = \sum_{k<l} a_{\psi(k)\psi(l)} = \sum_{k=1}^{n-1}\sum_{l=k+1}^{n} a_{\psi(k)\psi(l)} . \qquad (8.1)$$

We define *perfect dominance* as the condition when every element above the main diagonal is greater than or equal to its "mirror" below the diagonal, i.e., $l > k \Rightarrow a_{\psi(k)\psi(l)} \geq a_{\psi(l)\psi(k)}$. For a given matrix, **A**, with a given permutation of n objects, the algorithmic pseudocode for the EVALUATION of the dominance index is simple and inserts easily into the general branch-and-bound algorithmic pseudocode found in Appendix B.

DomInd = 0
for i = 1 to n − 1
   for j = i + 1 to n

DomInd = DomInd + **A**(*permutation*(i), *permutation*(j))
        next j
next i

8.2 Fathoming Tests for Optimizing the Dominance Index

8.2.1 Determining an Initial Lower Bound

Often, an initial lower bound in the branch-and-bound algorithm is set to 0. However, because the branch-and-bound technique is more efficient when sequences are pruned early, setting a more stringent lower bound is prudent. Moreover, a dominance index is not necessarily non-negative, making an initial lower bound of 0 overly presumptive. With the tendency to place target i to the left of target j in the sequence if $a_{ij} > a_{ji}$, the process of optimizing the dominance index is aided by an initial bound found by evaluating a permutation that orders objects according to their respective *rowsums*. The *rowsums* are useful in many fathoming operations. The most useful *rowsums* to store are those that ignore the diagonal, as though the diagonal entries were 0. Therefore, they should be calculated (with a 0 diagonal) and stored prior to implementation of the branch-and-bound algorithm. The mathematical model is stated formally below:

$$r_i = \sum_{j=1, j \neq i}^{n} a_{ij}, \text{ for } 1 \leq i \leq n. \tag{8.2}$$

We utilize a function that ascertains the position of an object in the highest-to-lowest permutation:

$$rank(i) = [\sum_{j=1}^{n} index_i(r_j)] + 1, \tag{8.3}$$

$$\text{where } index_i(r_j) = \begin{cases} 1 & \text{if } r_j > r_i \\ 1 & \text{if } r_j = r_i, j < i \\ 0 & \text{otherwise.} \end{cases} \tag{8.3.1}$$

8.2 Fathoming Tests for Optimizing the Dominance Index

The algorithmic psuedocode for finding the appropriate initial lower bound orders ranking ties in order of occurrence in the original (identity) permutation/matrix. This is a convention for breaking ties.

```
/* Calculate rowsums disregarding the main diagonal */
for i = 1 to n
    rowsum(i) = 0
    for j = 1 to n
        if i <> j then rowsum(i) = rowsum(i) + A(i, j)
    next j
next i
/* Rank order the rowsums into an initial permutation */
for i = 1 to n
    Set Index = 0 and EarlyTies = 0
    for j = 1 to n
        if rowsum(j) > rowsum(i) then Index = Index + 1
        if (rowsum(j) = rowsum(i) and j < i) Then EarlyTies = EarlyTies + 1
    next j
    permutation(Index + EarlyTies + 1) = i
next i
LowerB = EVALUATION
```

To bolster the initial lower bound, we can order the permutation as per maximum graduated *rowsums*. The maximum graduated *rowsums* are determined by subtracting the column entry of every object as it is chosen from the *rowsums* of unselected objects.

```
/* Initialize the permutation and the graduated rowsums */
for i = 1 to n
    permutation(i) = i
    GradSum(i) = rowsum(i)
next i
/* Find the maximum graduated rowsum of unselected objects */
for i = 1 To n
    Max = GradSum(permutation(i)) and Index = i
    for j = i + 1 To n
        if Max < GradSum(permutation(j)) then
            Max = GradSum(permutation(j)) and Index = j
    Next j
    /*Reorder the permutation to put the Max GradSum in the ith position */
    hold = permutation(Index)
    for k = i to Index
```

```
    hold2 = permutation(k), permutation(k) = hold, hold = hold2
  next k
  /* Adjust rowsums; remember that permutation(Index) is now
  permutation(i). */
  for j = i + 1 to n
    GradSum(permutation(j)) = GradSum(permutation(j))
                            - A(permutation(j), permutation(i))
  next j
next i
```

8.2.2 The Adjacency Test

The adjacency test for the dominance index amounts to the comparison of $a_{\psi(p-1)\psi(p)}$ to $a_{\psi(p)\psi(p-1)}$. By using this adjacency test, we are guaranteed to find an optimal permutation that satisfies the well-known necessary condition for optimal solutions known as the Hamiltonian ranking property, which is,

$$a_{\psi(k)\psi(k+1)} \geq a_{\psi(k+1)\psi(k)} \text{ for } 1 \leq k \leq n-1. \tag{8.4}$$

Specifically, if $a_{\psi(p-1)\psi(p)} \geq a_{\psi(p)\psi(p-1)}$ then the adjacency test pases. In terms of the pseudocode, if **A**(*permutation*(*Position* − 1), *permutation*(*Position*)) >= **A**(*permutation*(*Position*), *permutation*(*Position* − 1)) then ADJACENCYTEST = True. Otherwise, the adjacency test fails.

8.2.3 The Bound Test

To find an upper bound for a given partial ordering of n objects, we need to use some set notation. If we know our position in the sequence, then we know that $\psi(1), \psi(2), \ldots, \psi(p)$ have been assigned objects. This subset of objects is denoted as $R(p) = \{\psi(1), \psi(2), \ldots, \psi(p)\}$. For the dominance index, the entries to the right of the diagonal for every row are summed. With the known *rowsums* and known selected objects, we can quickly find the partial solution for the partial sequence.

```
PartSolution = 0
for i = 1 to Position
  PartSolution = PartSolution + rowsum(permutation(i))
  for j = 1 to i − 1
    PartSolution = PartSolution − A(permutation(i), permutation(j))
```

next j
next i

In mathematical parlance, the first component of the upper bound for a partial sequence in optimizing the dominance index is rather straightforward.

$$f_{B1} = \sum_{k=1}^{p} \left(r_{\psi(k)} - \sum_{l=1}^{k-1} a_{\psi(k)\psi(l)} \right). \tag{8.5.1}$$

The second component of the upper bound is concerned with the complement of $R(p)$ in the set of all n objects, $S\setminus R(p)$, which contains the remaining $(n - p)$ objects. Because the dominance index sums row entries to the right of the diagonal, we want to find the largest possible entries to the right of the diagonal in the rows for the remaining objects. We allow the complement to contain indices for the remaining objects with an initial index of *Position* + 1 for the sake of clarity.

```
/* Determine S\R(Position) */
index = Position + 1
for i = 1 to n
   found = False
   for j = 1 to Position
      if permutation(j) = i then found = true
   next j
   if not found then
      complement(index) = i
      index = index + 1
   end if
next i
/* Find maximum contribution to objective function value by the
complement */
ComplementContribution = 0
for i = Position + 1 to n − 1
   for j = i + 1 to n
      if A(complement(i), complement(j)) >
                       A(complement(j), complement(i)) then
         max = A(complement(i), complement(j))
      else
         max = A(complement(j), complement(i))
      end if
      ComplementContribution = ComplementContribution + max
```

8 Seriation—Maximization of a Dominance Index

next j
next i

With the understanding of how to calculate the highest possible contribution of $S\backslash R(Position)$ to the dominance index, the mathematical notation is succinct and sensible:

$$f_{B2} = \sum_{i<j \in S\backslash R(p)} \max(a_{ij}, a_{ji}). \qquad (8.5.2)$$

Thus, the upper bound for a partial sequence in optimizing the dominance index is $UpperB$ = PartSolution + ComplementContribution or, mathematically, $f_B = f_{B1} + f_{B2}$. If the upper bound, $UpperB$, is less than or equal to the previously determined lower bound, $LowerB$, then the bound test fails, i.e. BOUNDTEST = False.

8.3 Demonstrating the Iterative Process

To illustrate the process of the main branch-and-bound algorithm using the fathoming rules, we have constructed a 5 × 5 asymmetric matrix for which we can find the maximum dominance index. The matrix is shown in Table 8.1 and the initial lower bound for finding the optimal dominance index is determined as $f(\psi) = 52$ with $\psi = (3, 2, 5, 1, 4)$.

Table 8.1. A 5 × 5 matrix and a reordering of that matrix.

	5 × 5 Matrix						Reordered Matrix				
	1	2	3	4	5		3	2	5	1	4
1		5	4	5	1	3		7	3	4	8
2	3		1	7	6	2	1		6	3	7
3	4	7		8	3	5	5	2		7	2
4	3	1	0		6	1	4	5	1		5
5	7	2	5	2		4	0	1	6	3	

The reordered matrix gives us an initial lower bound of 52. Using this lower bound, we begin the main algorithm. The execution of the branch-and-bound algorithm shows beneficial pruning for many partial sequences, leading to right branches or retraction as shown in Table 8.2. As a note of interest, the number of iterations reduces from 22 to 17 by performing the branch-and-bound algorithm on the *re-ordered* matrix in Table 8.1. Re-ordering a matrix during the INITIALIZE step is a commonly used tactic to reduce computation time by encouraging the discovery of tighter upper bounds early in the algorithm.

Table 8.2. Finding the maximum dominance index for the matrix in Table 8.1.

Row	Partial sequence	Adjacency test	Partial solution	Maximum complement contribution	UpperB	Dispensation
1	1 - 2	Pass	29	19	48	Prune, 48 <= 52
2	1 - 3	Pass	33	19	52	Prune, 52 <= 52
3	1 - 4	Pass	22	18	40	Prune, 40 <= 52
4	1 - 5	Fail				Prune
5	1 - 6					Retraction
6	2	Pass	17	35	52	Prune, 52 <= 52
7	3	Pass	22	36	58	Branch Forward
8	3 - 1	Pass	33	19	52	Prune, 52 <= 52
9	3 - 2	Pass	38	18	56	Branch Forward
10	3 - 2 - 1	Fail				Prune
11	3 - 2 - 4	Pass	47	7	54	Branch Forward
12	3 - 2 - 4 - 1					Suboptimal, 48 <= 52
13	3 - 2 - 4 - 5					*New incumbent, $f^* = 54$
14	3 - 2 - 4 - 6					Retraction
15	3 - 2 - 5	Pass	47	5	52	Prune, 52 <= 54
16	3 - 2 - 6					Retraction
17	3 - 4	Pass	32	18	50	Prune, 50 <= 54
18	3 - 5	Fail				Prune
19	3 - 6					Retraction
20	4	Pass	10	34	44	Prune, 44 <= 54
21	5	Pass	16	36	52	Prune, 52 <= 54
22	6					

8.4 EXAMPLES—Extracting and Ordering a Subset

8.4.1 Tournament Matrices

One interesting problem for maximizing the dominance index corresponds to a type of paired-comparison matrix as generated by a tournament. (As an additional note, tournament matrices have binary entries and exhibit the characteristic of $a_{ij} + a_{ji} = 1$ for all $i \neq j$. Not all paired-comparison matrices have binary entries.) Maximizing the dominance index produces a permutation that tends to place the 1s in the upper triangle (and 0s generally in the lower triangle) of the re-ordered matrix. As an example, we refer to a 15 × 15 tournament matrix originally reported by Hubert and Schultz (1975).

Table 8.3. A 15 × 15 tournament matrix (Hubert & Schultz, 1975).

	1	2	3	4	5	6	7	8	9	10	11	12	13	14	15
1	0	1	1	1	0	0	0	1	1	0	0	0	1	0	0
2	0	0	1	0	0	0	0	0	1	1	1	1	1	1	0
3	0	0	0	1	1	0	0	1	0	0	1	0	1	0	1
4	0	1	0	0	0	0	0	0	0	1	0	1	1	0	0
5	1	1	0	1	0	1	1	1	0	0	0	1	1	0	1
6	1	1	1	1	0	0	1	1	0	1	1	0	1	1	1
7	1	1	1	1	0	0	0	1	1	0	0	0	1	1	1
8	0	1	0	1	0	0	0	0	1	0	0	0	1	0	1
9	0	0	1	1	1	1	0	0	0	1	0	0	0	1	0
10	1	0	1	0	1	0	1	1	0	0	0	1	0	0	1
11	1	0	0	1	1	0	1	1	1	1	0	1	1	1	1
12	1	0	1	0	0	1	1	1	1	0	0	0	1	0	0
13	0	0	0	0	0	0	0	0	1	1	0	0	0	0	1
14	1	0	1	1	1	0	0	1	0	1	0	1	1	0	1
15	1	1	0	1	0	0	0	0	1	0	0	1	0	0	0

The columns and rows can be permuted to attain the maximum dominance index of $f^*(\psi) = 83$ with the optimal permutation of $\psi = (6, 2, 11, 14, 10, 5, 12, 7, 1, 3, 8, 13, 15, 9, 4)$ as shown in Table 8.4. Clearly, although the 1s tend to be above the diagonal, this matrix does not exhibit perfect dominance. An extension of the problem of maximizing the dominance index is to find a subset with perfect dominance. In the case of a tournament matrix, this will be a submatrix with all of the 1s in the upper triangle. To achieve this end, we modify our algorithm to continue using n as the number of objects but restrict our EVALUATION criteria to the subset size under consideration. If no such subset can be found, then we decrement the subset size and continue. In this way, we can find the largest subset size with perfect dominance. For the matrix in Table 8.3, the maximum subset size that can attain perfect dominance is 8. In this example, that objective value is $(8^2 - 8)/2 = 28$ with the permutation of $\psi = (5, 6, 7, 1, 8, 4, 2, 13)$.

Table 8.4. A 15 × 15 tournament matrix from Hubert & Schultz (1975) permuted to achieve a maximum dominance index.

	6	2	11	14	10	5	12	7	1	3	8	13	15	9	4
6	0	1	1	1	1	0	0	1	1	1	1	1	1	0	1
2	0	0	1	1	1	0	1	0	0	1	0	1	0	1	0
11	0	0	0	1	1	1	1	1	1	0	1	1	1	1	1
14	0	0	0	0	1	1	1	0	1	1	1	1	1	0	1
10	0	0	0	0	0	1	1	1	1	1	1	0	1	0	0
5	1	1	0	0	0	0	1	1	1	0	1	1	1	0	1
12	1	0	0	0	0	0	0	1	1	1	1	1	0	1	0
7	0	1	0	1	0	0	0	0	1	1	1	1	1	1	1
1	0	1	0	0	0	0	0	0	0	1	1	1	0	1	1
3	0	0	1	0	0	1	0	0	0	0	1	1	1	0	1
8	0	1	0	0	0	0	0	0	0	0	0	1	1	1	1
13	0	0	0	0	1	0	0	0	0	0	0	0	1	1	0
15	0	1	0	0	0	0	1	0	1	0	0	0	0	1	1
9	1	0	0	1	1	1	0	0	0	1	0	0	0	0	1
4	0	1	0	0	1	0	1	0	0	0	0	1	0	0	0

Table 8.5. An 8 × 8 tournament submatrix exhibiting perfect dominance.

	5	6	7	1	8	4	2	13
5	0	1	1	1	1	1	1	1
6	0	0	1	1	1	1	1	1
7	0	0	0	1	1	1	1	1
1	0	0	0	0	1	1	1	1
8	0	0	0	0	0	1	1	1
4	0	0	0	0	0	0	1	1
2	0	0	0	0	0	0	0	1
13	0	0	0	0	0	0	0	0

The reason that this is an especially interesting case is that maximizing the dominance index will produce a permutation to yield perfect dominance *whenever possible* for a tournament matrix, placing the highest possible value (1) in all of the matrix entries in the upper triangle. Thus, if a tournament matrix has a submatrix exhibiting perfect dominance, then the submatrix will definitely maximize the dominance index for all subsets of that size. However, for many matrices, maximizing the dominance index for some subset of $n_S < n$ objects does not necessarily imply finding perfect dominance, even if this goal can be achieved for a different subset with n_S objects. This is true for other paired-comparison matrices recording proportions with the property that $a_{ij} + a_{ji} = c$, i.e. the mirror entries sum to a constant. If a paired comparison is divided by this constant, then the mirror entries sum to 1.

8.4.2 Maximum Dominance Index vs. Perfect Dominance for Subsets

An extension of the previous exercise is to determine whether or not we can find a subset with perfect dominance and, if so, maximize the dominance index for subsets of the same size. The fathoming tests are reduced to ensuring that the Hamiltonian ranking condition is met for all matrix entries in the new column (*Position*). The lower bound is not used. The main algorithm becomes:

Set *MaxSize* = 0
Set *Position* = 1 and *permutation*(*Position*) = 1
for k = 2 to *n*
   permutation (k) = 0
next k
while (*Position* <> 1 or *permutation*(*Position*) <= *n*)
   Position = *Position* + 1
   Perfect = False
   while not *Perfect*
      permutation(*Position*) = *permutation*(*Position*) + 1
      NotRedundancy = True
      for k = 1 to *Position* – 1
         if *permutation*(*Position*) = *permutation*(k) then
            NotRedundancy = False
      next k
      if *NotRedundancy* then
         if (*Position* = 1 and *permutation*(*Position*) > *n*) then exit loop
         if (*Position* > 1 and *permutation*(*Position*) > *n*) then
            permutation(*Position*) = 0
            Position = *Position* – 1
         else
            Perfect = Hamiltonian
            if *Perfect* and *Position* > *MaxSize* then
                BestSolution = *permutation*
                MaxSize = *Position*
                DominanceIndex = EVALUATION
            end if
         end if
      end if   /* No Redundancy */

8.4 EXAMPLES—Extracting and Ordering a Subset

   loop   /* Perfect loop */
loop   /* Termination loop */

The EVALUATION routine is slightly modified to evaluate the submatrix using *MaxSize* (in lieu of *n*) as the current number of objects.

EVALUATION = 0
for i = 1 to *MaxSize* − 1
  for j = i + 1 to *MaxSize*
    EVALUATION = EVALUATION + **A**(*permutation*(i), *permutation*(j))
  next j
next i

For the specialized case of finding a submatrix with perfect dominance, we employ a subroutine to check the Hamiltonian condition, which takes the place of the fathoming routines:

Hamiltonian = True
for i = 1 to *Position* − 1
  if **A**(*permutation*(i), *permutation*(*Position*))
                       < **A**(*permutation*(*Position*), *permutation*(i)) then
    Hamiltonian = False
  end if
next i

Once the maximum size for a subset with perfect dominance has been found, we can find a maximum dominance index for the *MaxSize* using all *n* objects. The fathoming tests should be modified accordingly. Instead of using the bound test, we can extend the adjacency test to an interchange test.

To extend the adjacency test to an interchange test, consider swapping objects *p* and *q* < *p*. The change in the contribution to the objective function is determined by the differences in which matrix entries from the rows/columns of *p* and *q* will be in the upper triangle. Specifically, in the given partial sequence, the entries on row *q* from the diagonal to position *p* and the entries on column *p* from position *q* to the diagonal will be summed in the objective function (SumRight). However, if the objects in positions *p* and *q* are exchanged, then this arrangement is reversed.

/* **Initialize** SumRight for the given partial permutation, SumRight(Current), and SumRight that will result from changing the permutation by swapping the object in *Position* with one of the previously chosen objects, SumRight(Swap). */
Set SumRight(Current) = 0 and SumRight(Swap) = 0.

```
for i = AltPosition + 1 to Position
  SumRight(Current) = SumRight(Current)
                      + A(permutation(AltPosition), permutation(i))
  SumRight(Swap) = SumRight(Swap)
                      + A(permutation (i), permutation(AltPosition))
next i
for j = AltPosition to Position − 1
  SumRight(Current) = SumRight(Current)
                      + A(permutation (j), permutation(Position))
  SumRight(Swap) = SumRight(Swap)
                      + A(permutation(Position), permutation(i))
next j
```

If the SumRight for any swap (i.e., for any AltPosition < *Postion*) results in a higher sum, then the INTERCHANGETEST fails. Because the adjacency test is a simple and quick check to determine the feasibility of the current partial sequence, we can quickly check our sequence using ADJACENCYTEST before invoking the more computationally demanding INTERCHANGETEST. Now, our fathoming rules read, "if ADJACENCYTEST then fathom = INTERCHANGETEST".

We applied this specialized branch-and-bound procedure to the 15 × 15 paired-comparison matrix in Table 8.6, which concerns the severity of criminal offenses (Thurstone, 1927). The data matrix in Table 8.6 was also published by Hubert and Golledge (1981, p. 435), who used the data to demonstrate the dynamic programming paradigm for maximizing the dominance index. The matrix rows and columns can be permuted to attain the maximum dominance index of $f^*(\psi) = 78946$ with the optimal permutation of $\psi =$ (12, 11, 4, 5, 8, 6, 13, 7, 3, 10, 2, 15, 1, 14, 9). However, the maximum size of a subset of the 15 objects to achieve perfect dominance is 13 with the permutation of $\psi =$ (12, 11, 4, 5, 8, 6, 3, 10, 2, 15, 1, 14, 9), which yields a dominance index of 60096 (see Table 8.7). Yet, as shown in Table 8.8, the maximum dominance index for a subset of size 13 is $f^*(\psi) = 60387$ with the permutation of $\psi =$ (12, 11, 5, 8, 6, 7, 3, 10, 2, 15, 1, 14, 9). The primary purpose of these examples is to demonstrate the ease with which branch-and-bound algorithms can be manipulated to accommodate seriation problems. Modifications are simply made to the routines to produce the desired modeling alternatives associated with the task at hand.

Table 8.6. A 15 × 15 matrix from Thurstone (1927) recording proportions of people rating severity of criminal offenses. The criminal offenses are Abortion (1), Adultery (2), Arson (3), Assault & Battery (4), Burglary (5), Counterfeiting (6), Embezzlement (7), Forgery (8), Homicide (9), Kidnapping (10), Larceny (11), Libel (12), Perjury (13), Rape (14), Seduction (15).

	1	2	3	4	5	6	7	8	9	10	11	12	13	14	15
1	0	323	338	211	238	244	245	212	760	318	222	191	256	822	419
2	677	0	415	242	281	285	253	274	863	365	207	182	245	925	589
3	662	585	0	260	226	321	348	254	917	563	215	144	349	944	716
4	789	757	740	0	515	556	485	534	970	743	385	385	587	947	785
5	762	719	774	485	0	593	605	580	981	856	333	322	478	981	769
6	756	715	679	444	407	0	540	488	947	804	303	284	532	963	756
7	755	747	652	515	395	460	0	350	958	752	305	248	474	977	774
8	788	726	746	466	420	512	650	0	951	819	343	320	534	966	820
9	240	137	83	30	19	53	42	49	0	83	30	34	79	441	181
10	682	635	437	257	144	196	248	181	917	0	170	106	288	902	595
11	778	793	785	615	667	697	695	657	970	830	0	348	648	970	848
12	809	818	855	615	678	716	752	680	966	894	652	0	702	981	886
13	744	755	651	413	522	467	526	466	921	712	352	298	0	951	767
14	178	75	56	53	19	37	23	34	559	98	30	19	49	0	76
15	581	411	284	215	231	244	226	180	819	405	152	114	233	924	0

Table 8.7. A 13 × 13 submatrix extracted from Thurstone's (1927) data and exhibiting perfect dominance.

	12	11	4	5	8	6	3	10	2	15	1	14	9
12	0	652	615	678	680	716	855	894	818	886	809	981	966
11	348	0	615	667	657	697	785	830	793	848	778	970	970
4	385	385	0	515	534	556	740	743	757	785	789	947	970
5	322	333	485	0	580	593	774	856	719	769	762	981	981
8	320	343	466	420	0	512	746	819	726	820	788	966	951
6	284	303	444	407	488	0	679	804	715	756	756	963	947
3	144	215	260	226	254	321	0	563	585	716	662	944	917
10	106	170	257	144	181	196	437	0	635	595	682	902	917
2	182	207	242	281	274	285	415	365	0	589	677	925	863
15	114	152	215	231	180	244	284	405	411	0	581	924	819
1	191	222	211	238	212	244	338	318	323	419	0	822	760
14	19	30	53	19	34	37	56	98	75	76	178	0	559
9	34	30	30	19	49	53	83	83	137	181	240	441	0

Table 8.8. A 13 × 13 submatrix extracted from Thurstone (1927) with a maximum dominance index for all subsets of size 13.

	12	11	5	8	6	7	3	10	2	15	1	14	9
12	0	652	678	680	716	752	855	894	818	886	809	981	966
11	348	0	667	657	697	695	785	830	793	848	778	970	970
5	322	333	0	580	593	605	774	856	719	769	762	981	981
8	320	343	420	0	512	650	746	819	726	820	788	966	951
6	284	303	407	488	0	540	679	804	715	756	756	963	947
7	248	305	395	350	460	0	652	752	747	774	755	977	958
3	144	215	226	254	321	348	0	563	585	716	662	944	917
10	106	170	144	181	196	248	437	0	635	595	682	902	917
2	182	207	281	274	285	253	415	365	0	589	677	925	863
15	114	152	231	180	244	226	284	405	411	0	581	924	819
1	191	222	238	212	244	245	338	318	323	419	0	822	760
14	19	30	19	34	37	23	56	98	75	76	178	0	559
9	34	30	19	49	53	42	83	83	137	181	240	441	0

8.5 Strengths and Limitations

Branch-and-bound algorithms for maximizing the dominance index of matrices are straightforward. This attribute of the methodology lends itself to ease of adaptation to specific problems, as demonstrated in the previous section. Hence, the breadth of applicability of branch-and-bound procedures is quite large for these problems.

However, the algorithm can be sensitive to the characteristics of the proximity matrix being examined. Some matrices with as few as 20 objects can require significant computation time, particularly when there is a serious departure from perfect dominance and/or the elements of the proximity matrix have a small range. For example, randomly generated binary matrices can be particularly difficult to solve. Dynamic programming is a more reliable (i.e., less sensitive to the structure of the matrix) strategy than branch-and-bound when computer memory permits its implementation. For large proximity matrices, a branch-and-bound integer programming approach might be the most fruitful strategy.

8.6 Available Software

We have made available a software program for finding a permutation of the rows and columns of an asymmetric matrix so as to maximize the dominance index. This program, *bbdom.for*, can be accessed and downloaded from either http://www.psiheart.net/quantpsych/monograph.html or http://garnet.acns.fsu.edu/~mbrusco. This program reads an $n \times n$ nonnegative asymmetric matrix from the file *asym.dat*. The first element of the file must be the number of objects, n, and the remaining data in the file are the elements of the asymmetric matrix. The program writes output to the file *asym.out*. The output consists of the index value for the initial lower bound, the optimal dominance index value, the required CPU time, and the optimal permutation.

The first phase of *bbdom.for* uses a pairwise interchange heuristic to establish a lower bound for the dominance index and to provide a reordering of the objects. Specifically, a permutation of objects is randomly generated and pairwise interchange operations are applied until no such local search operation will improve the dominance index value. This process is repeated using 100 different initial random permutations and the best solution across the 100 trials is stored. The best permutation is then used to reorder the objects prior to entering the branch-and-bound process.

The branch-and-bound code contains an adjacency test, an interchange test, a single-object relocation test, and a bound test for pruning partial solutions. A single object relocation test is very similar to an interchange test. However, a single-object relocation test considers the effect of placing the object in position p into position q and pushing all other objects to the next position from their current position in the permutation, rather than swapping positions of the objects in positions p and q. Any of these tests can be bypassed using a simple modification of the code. In fact, we have found the interchange test and single-object relocation test provide approximately equal benefit with respect to pruning. However, because the relocation test is slightly more efficient with respect to its evaluation, its deployment in favor of the interchange test is recommended.

The application of *bbdom.for* to the following matrices from the empirical literature produces the following results:

Results for the data in Table 8.6, Thurstone's (1927) severity of criminal offenses data from Hubert and Golledge (1981, p. 435):
MAXIMUM DOMINANCE INDEX   78946.0000 CPU TIME   0.01

12 11 4 5 8 6 13 7 3 10 2 15 1 14 9

Results for Rodgers and Thompson's (1992, p. 113) co-citation matrix among past presidents of the Psychometric Society:
MAXIMUM DOMINANCE INDEX   499.0000 CPU TIME   0.02
18 17 13 9 11 6 12 14 5 4 16 15 8 2 3 10 1 7

Results for van der Heijden, Malthas, and van den Roovaart's (1984) interletter confusion matrix from Heiser (1988, p. 41):
MAXIMUM DOMINANCE INDEX   10577.0000 CPU TIME   0.90
19 5 22 6 12 21 23 13 2 1 8 14 11 24 18 17 15 3 7 4 20 25 9 16 26 10

Results for Groenen and Heiser's (1996, p. 547) cross-citation matrix among journals:
MAXIMUM DOMINANCE INDEX   10193.0000 CPU TIME   0.14
19 8 5 14 26 6 7 11 1 22 24 18 3 16 12 15 21 27 9 13 20 4 25 28 10 17 2 23

The CPU times for many of these matrices are quite modest. To some extent, CPU times are a function of problem size, but note that the 26 × 26 matrix required more time than the 28 × 28 matrix. The algorithm is, in fact, quite sensitive to the matrix properties. For example, some binary matrices of modest size (e.g., 25 × 25) might require an hour or more of CPU time.

9 Seriation—Maximization of Gradient Indices

9.1 Introduction to the Gradient Indices

Gradients within matrices are best understood as a flow of "greater than" or "less than" relationships between matrix entries. These indices reveal structure (or lack thereof) within matrices such as Robinson structure (Robinson, 1951) within a symmetric dissimilarity matrix. More to the point, anti-Robinson patterning is essential for various seriation objectives. Anti-Robinson structure is found when matrix entries are nondecreasing when moving away from the diagonal in the rows and columns, pushing larger entries away from the diagonal and, hence, more dissimilar objects away from each other (i.e., toward opposite ends of the sequence).

The importance of Robinson and anti-Robinson structure is well-documented in the seriation literature (Hubert, 1987; Hubert & Arabie, 1994; Hubert, Arabie, & Meulman, 1998). For example, there are streams of research that focus on the graphical representation of proximity data exhibiting a Robinson or anti-Robinson structure (Diday, 1986; Durand & Fichet, 1988), as well the fitting of proximity matrices via sums of matrices associated with Robinson or anti-Robinson form (Hubert & Arabie, 1994; Hubert et al., 1998). Throughout the remainder of this chapter, however, we will limit our interest to finding permutations for a single dissimilarity matrix based on anti-Robinson structure.

Hubert et al. (2001, Chapter 4) describe several within-row and/or within-column gradient indices designed to induce anti-Robinson patterning in a symmetric dissimilarity matrix by reordering the matrix. There are four gradient indices of interest—unweighted within row gradient, unweighted within row and column gradient, weighted within row gradient, weighted within row and column gradient. The unweighted row and column gradients tally the sign of the differences between entries, whereas the weighted row and column gradients tally the actual raw differences between entries in the rows/columns. A useful function for these gradients is:

9 Seriation—Maximization of Gradient Indices

$$sign(x) = \begin{cases} +1 & \text{if } x > 0 \\ 0 & \text{if } x = 0 \\ -1 & \text{if } x < 0. \end{cases} \quad (9.1)$$

Mathematically, we use equation (9.1) and keep our indices—i, j, and k—distinct to order the matrix to maximize (greater than) or minimize (less than) the index. We can state the unweighted within row gradient as:

$$U_r = \sum_{i=1}^{n-2} \sum_{j=i+1}^{n-1} \sum_{k=j+1}^{n} sign(a_{\psi(i)\psi(k)} - a_{\psi(i)\psi(j)}), \quad (9.2.1)$$

and the unweighted within row and column gradient as:

$$U_{rc} = \sum_{i=1}^{n-2} \sum_{j=i+1}^{n-1} \sum_{k=j+1}^{n} (sign(a_{\psi(i)\psi(k)} - a_{\psi(i)\psi(j)}) + sign(a_{\psi(i)\psi(k)} - a_{\psi(j)\psi(k)}). \quad (9.2.2)$$

Maximizing equation (9.2.2) is equivalent to maximizing the anti-Robinson form of the symmetric matrix. The algorithmic pseudocode for the unweighted gradient within rows *and* columns uses conditional statements to tally the index of interest.

Unweighted = 0
for i = 1 to n – 2
  for j = i + 1 to n – 1
    for k = j + 1 to n
      difference = **A**(*permutation*(i), *permutation*(k)) –
                                      – **A**(*permutation*(i), *permutation*(j))
      /* Tally for rows */
      if difference > 0 then Unweighted = Unweighted + 1
      if difference < 0 then Unweighted = Unweighted – 1
      difference = **A**(*permutation*(i), *permutation*(k))
                                      – **A**(*permutation*(j), *permutation*(k))
      /* Tally for columns */
      if difference > 0 then Unweighted = Unweighted + 1
      if difference < 0 then Unweighted = Unweighted – 1
    next k
  next j
next i

The mathematical and algorithmic representations of the weighted gradients for rows and columns do not require the conditional statements but do require some computational effort. The weighted within row gradient is

$$W_r = \sum_{i=1}^{n-2} \sum_{j=i+1}^{n-1} \sum_{k=j+1}^{n} (a_{\psi(i)\psi(k)} - a_{\psi(i)\psi(j)}), \qquad (9.2.3)$$

whereas the weighted within row and column gradient is

$$W_{rc} = \sum_{i=1}^{n-2} \sum_{j=i+1}^{n-1} \sum_{k=j+1}^{n} (2a_{\psi(i)\psi(k)} - a_{\psi(i)\psi(j)} - a_{\psi(j)\psi(k)}). \qquad (9.2.4)$$

For the weighted within row and column gradient, the appropriate algorithmic pseudocode tallies 2*A(*permutation*(i), *permutation*(k)) – A(*permutation*(i), *permutation*(j)) – A(*permutation*(j), *permutation*(k)).

9.2 Fathoming Tests for Optimizing Gradient Indices

9.2.1 The Initial Lower Bounds for Gradient Indices

The initial lower bound for maximizing a gradient index is determined in a three-step process. First, we rank the rows in terms of graduated rowsums as we did with maximizing the dominance index. However, because we are dealing with symmetric matrices, we then re-order the ranks in an alternating fashion. Finally, we use an interchange heuristic to converge toward a local optimum. The initial lower bound is then simply the evaluation of the resulting permutation.

Once the permutation has been found to place the objects in order of their graduated *rowsums*, the permutation can be put into the *ends_rank* order by placing the highest ranks at the ends of the permutation and successive ranks alternating at the ends toward the middle. Maximizing gradient indices for symmetric matrices tends to put objects with higher entries at opposite ends of the sequence. Therefore, a prudent initial lower bound would be the evaluation of a permutation for which objects with higher graduated *rowsums* alternate at the ends of the sequence, working toward the middle of the sequence. Fortunately, we can use the highest-to-lowest ranking function of equations (8.3) and (8.3.1) followed by a reordering of the ranks to mimic the objective functions for gradient indices. In particular, we want a type of alternating ranking system for n objects.

$$ends\_rank(i) = \begin{cases} rank(2i-1) & , i < (n/2)+1 \\ rank(2*(n-i+1)) & , \text{otherwise}. \end{cases} \quad (9.3)$$

Algorithmically, we find the permutation of alternating ranks with two simple conditional statements.

for i = 1 to n
  if i < n/2 + 1 then *permutation*(i) = *rank*(2i − 1)
  if i >= n/2 + 1 then *permutation*(i) = *rank*(n − i + 1)
next i

With the alternating graduated *rowsums* in place, we use a pairwise interchange heuristic to quickly approach a local optimum.

for AltPosition = 1 to n − 1
 CurrentContribution = EVALUATION
  for LaterPosition = AltPosition + 1 to n
  /* Perform **EXCHANGETEST** for AltPosition and LaterPosition. */
  hold = permutation(LaterPosition)
  permutation(LaterPosition) = permutation(AltPosition)
  permutation(AltPosition) = hold
  AltContribution = EVALUATION
  if CurrentContribution < AltContribution then
   CurrentContribution = AltContribution
  else
   hold = *permutation*(LaterPosition)
   permutation(LaterPosition) = *permutation*(AltPosition)
   permutation(AltPosition) = hold
  end if
 next LaterPosition
next AltPosition

9.2.2 The Adjacency Test for Gradient Indices

For maximizing gradient indices, the adjacency test is a powerful fathoming instrument. When comparing contributions to the objective function value by the position of an object, $\psi(p)$, and the preceding object, $\psi(p-1)$, all relevant triplets involving both p and $p-1$ must be evaluated. The precise adjacency test for each gradient index varies slightly

9.2 Fathoming Tests for Optimizing Gradient Indices 117

from the others but holds to a roughly similar pattern. The adjacency tests for criteria involving column examinations require the formation of a set, $S\backslash R(p)$, containing all unassigned objects, i.e. the complement in S of the set of objects chosen for the first p positions in the sequence, as was used in the bound test for maximizing the dominance index.

The following equations were developed (Brusco, 2002b) as successful adjacency tests for the corresponding gradient indices of equations (9.2.1) through (9.2.4). The left-hand side of each equation totals the anti-Robinson index contributions to the objective function for the criterion by $\psi(p-1)$ and $\psi(p)$ when $\psi(p-1)$ precedes $\psi(p)$; the right-hand side represents contributions if $\psi(p-1)$ and $\psi(p)$ exchange places in the sequence.

$$\left(\sum_{i=1}^{p-2} sign(a_{\psi(i)\psi(p)} - a_{\psi(i)\psi(p-1)})\right) + \left(\sum_{j \notin R(p)} sign(a_{\psi(p-1)j} - a_{\psi(p-1)\psi(p)})\right) \quad (9.4.1)$$

$$\geq \left(\sum_{i=1}^{p-2} sign(a_{\psi(i)\psi(p-1)} - a_{\psi(i)\psi(p)})\right) + \left(\sum_{j \notin R(p)} sign(a_{\psi(p)j} - a_{\psi(p)\psi(p-1)})\right)$$

$$\left(\sum_{i=1}^{p-2}(sign(a_{\psi(i)\psi(p)} - a_{\psi(i)\psi(p-1)}) + sign(a_{\psi(i)\psi(p)} - a_{\psi(p-1)\psi(p)}))\right) \quad (9.4.2)$$

$$+ \left(\sum_{j \notin R(p)}(sign(a_{\psi(p-1)j} - a_{\psi(p-1)\psi(p)}) + sign(a_{\psi(p-1)j} - a_{\psi(p)j}))\right)$$

$$\geq \left(\sum_{i=1}^{p-2}(sign(a_{\psi(i)\psi(p-1)} - a_{\psi(i)\psi(p)}) + sign(a_{\psi(i)\psi(p-1)} - a_{\psi(p)\psi(p-1)}))\right)$$

$$+ \left(\sum_{j \notin R(p)}(sign(a_{\psi(p)j} - a_{\psi(p)\psi(p-1)}) + sign(a_{\psi(p)j} - a_{\psi(p-1)j}))\right)$$

$$\left(\sum_{i=1}^{p-2}(a_{\psi(i)\psi(p)} - a_{\psi(i)\psi(p-1)})\right) + \left(\sum_{j \notin R(p)}(a_{\psi(p-1)j} - a_{\psi(p-1)\psi(p)})\right) \quad (9.4.3)$$

$$\geq \left(\sum_{i=1}^{p-2}(a_{\psi(i)\psi(p-1)} - a_{\psi(i)\psi(p)})\right) + \left(\sum_{j \notin R(p)}(a_{\psi(p)j} - a_{\psi(p)\psi(p-1)})\right)$$

$$\left(\sum_{i=1}^{p-2} (2a_{\psi(i)\psi(p)} - a_{\psi(i)\psi(p-1)} - a_{\psi(p-1)\psi(p)}) \right) \quad (9.4.4)$$

$$+ \left(\sum_{j \notin R(p)} (2a_{\psi(p-1)j} - a_{\psi(p-1)\psi(p)} - a_{\psi(p)j}) \right)$$

$$\geq \left(\sum_{i=1}^{p-2} (2a_{\psi(i)\psi(p-1)} - a_{\psi(i)\psi(p)} - a_{\psi(p)\psi(p-1)}) \right)$$

$$+ \left(\sum_{j \notin R(p)} (2a_{\psi(p)j} - a_{\psi(p)\psi(p-1)} - a_{\psi(p-1)j}) \right)$$

These equations for the passing conditions of the adjacency tests can be reduced for ease of implementation.

$$\sum_{i=1}^{p-2} 2(sign(a_{\psi(i)\psi(p)} - a_{\psi(i)\psi(p-1)}) \quad (9.5.1)$$

$$\geq \sum_{j \notin R(p)} (sign(a_{\psi(p)j} - a_{\psi(p)\psi(p-1)}) - sign(a_{\psi(p-1)j} - a_{\psi(p-1)\psi(p)}))$$

$$\sum_{i=1}^{p-2} (2sign(a_{\psi(i)\psi(p)} - a_{\psi(i)\psi(p-1)}) \quad (9.5.2)$$

$$+ sign(a_{\psi(i)\psi(p)} - a_{\psi(p-1)\psi(p)})$$

$$- sign(a_{\psi(i)\psi(p-1)} - a_{\psi(p)\psi(p-1)}))$$

$$\geq \sum_{j \notin R(p)} (sign(a_{\psi(p)j} - a_{\psi(p)\psi(p-1)})$$

$$+ sign(a_{\psi(p)j} - a_{\psi(p-1)j})$$

$$- sign(a_{\psi(p-1)j} - a_{\psi(p-1)\psi(p)})$$

$$- sign(a_{\psi(p-1)j} - a_{\psi(p)j}))$$

$$\sum_{i=1}^{p-2} 2(a_{\psi(i)\psi(p)} - a_{\psi(i)\psi(p-1)}) \tag{9.5.3}$$
$$\geq \sum_{j \notin R(p)} (a_{\psi(p)j} - a_{\psi(p-1)j} + a_{\psi(p-1)\psi(p)} - a_{\psi(p)\psi(p-1)})$$

$$\sum_{i=1}^{p-2} (3a_{\psi(i)\psi(p)} - 3a_{\psi(i)\psi(p-1)} - a_{\psi(p-1)\psi(p)} + a_{\psi(p)\psi(p-1)}) \tag{9.5.4}$$
$$\geq \sum_{j \notin R(p)} (3a_{\psi(p)j} - 3a_{\psi(p-1)j} - a_{\psi(p)\psi(p-1)} + a_{\psi(p-1)\psi(p)})$$

Notice that $a_{\psi(p-1)\psi(p)} = a_{\psi(p)\psi(p-1)}$ in the adjacency test for the weighted within row gradient of a *symmetric* matrix, further reducing the computation. To demonstrate the ease of implementation, we can develop the pseudocode for the unweighted gradient for rows and columns.

```
/* Initialize the sides of the equation */
Set compare1 = 0 and compare2 = 0
/* Calculate left-hand side of the equation */
for i = 1 To Position – 2
    difference = A(permutation(i), permutation(Position))
                 – A(permutation(i), permutation(Position - 1))
    /* Subtract sign of difference if difference < 0 */
    if difference > 0 then compare1 = compare1 + 2
    if difference < 0 then compare1 = compare1 – 2
    difference = A(permutation(i), permutation(Position))
                 – A(permutation(Position - 1), permutation(Position))
    /* Subtract sign of difference if < 0 */
    if difference > 0 then compare1 = compare1 + 1
    if difference < 0 then compare1 = compare1 – 1
    difference = A(permutation(i), permutation(Position – 1))
                 – A(permutation(Position), permutation(Position – 1))
    /* Subtract sign of difference if < 0 */
    if difference > 0 then compare1 = compare1 – 1
    if difference < 0 then compare1 = compare1 + 1
next i
/* Determine S\R(Position) */
Index = Position + 1
for i = 1 To n
```

```
   found = False
   for j = 1 To Position
      if permutation(j) = i then found = True
   next j
   if not found then
      complement(Index) = i
      Index = Index + 1
   end if
next i
/* Calculate right-hand side of the equation */
compare2 = 0
for j = Position + 1 to n
   difference = A(permutation(Position), complement(j))
                  − A(permutation(Position), permutation(Position − 1))
   /* Subtract sign of difference if < 0 */
   if difference > 0 then compare2 = compare2 + 1
   if difference < 0 then compare2 = compare2 − 1
   difference = A(permutation(Position), complement(j))
                  −A(permutation(Position − 1), complement(j))
   /* Subtract sign of difference if < 0 */
   if difference > 0 then compare2 = compare2 + 1
   if difference < 0 then compare2 = compare2 − 1
   difference = A(permutation(Position − 1), complement(j))
                  −A(permutation(Position − 1), permutation(Position))
   /* Subtract sign of difference if < 0 */
   if difference > 0 then compare2 = compare2 − 1
   if difference < 0 then compare2 = compare2 + 1
   difference = A(permutation(Position − 1), complement(j))
                  −A(permutation(Position), complement(j))
   /* Subtract sign of difference if < 0 */
   if difference > 0 then compare2 = compare2 − 1
   if difference < 0 then compare2 = compare2 + 1
next j
```

We simply compare the left-hand side of the equation with the right-hand side, *compare1* >= *compare2*, to determine whether or not the partial sequence passes the adjacency test.

9.2.3 The Bound Test for Gradient Indices

As with the adjacency tests, the bound tests for equations (9.2.1) through (9.2.4) are unique yet not too dissimilar. The upper bounds for the unweighted gradient indices incorporate a constant term, $b = p(n-p)(n-p-1)/2 + (n-p)(n-p-1)(n-p-2)/6$, to account for the number of terms corresponding to triples formed by one of the p objects in the partial ordering and pairs of distinct i, j not in the complement, $S|R(p)$, summed with the number of terms corresponding to triples formed by distinct i, j, k not in the complement (hence, this second term for the within row and column index is multiplied by 2, making the divisor equal to 3).

$$f_{Ur} = \left(\sum_{i=1}^{p-2} \sum_{j=i+1}^{p-1} \sum_{k=j+1}^{p} sign(a_{\psi(i)\psi(k)} - a_{\psi(i)\psi(j)}) \right) + \left(\sum_{i=1}^{p-1} \sum_{j=i+1}^{p} \sum_{k \notin R(p)} sign(a_{\psi(i)k} - a_{\psi(i)\psi(j)}) \right) \quad (9.6.1)$$
$$+ p(n-p)(n-p-1)/2 + (n-p)(n-p-1)(n-p-2)/6$$

$$f_{Urc} = \left(\sum_{i=1}^{p-2} \sum_{j=i+1}^{p-1} \sum_{k=j+1}^{p} (sign(a_{\psi(i)\psi(k)} - a_{\psi(i)\psi(j)}) + sign(a_{\psi(i)\psi(k)} - a_{\psi(j)\psi(k)})) \right) \quad (9.6.2)$$
$$+ \left(\sum_{i=1}^{p-1} \sum_{j=i+1}^{p} \sum_{k \notin R(p)} (sign(a_{\psi(i)k} - a_{\psi(i)\psi(j)}) + sign(a_{\psi(i)k} - a_{\psi(j)k})) \right)$$
$$+ p(n-p)(n-p-1) + (n-p)(n-p-1)(n-p-2)/3$$

$$f_{Wr} = \sum_{i=1}^{p-2} \sum_{j=i+1}^{p-1} \sum_{k=j+1}^{p} (a_{\psi(i)\psi(k)} - a_{\psi(i)\psi(j)}) + \sum_{i=1}^{p-1} \sum_{j=i+1}^{p} \sum_{k \notin R(p)} (a_{\psi(i)k} - a_{\psi(i)\psi(j)}) \quad (9.6.3)$$
$$+ \sum_{(i<j) \notin R(p)} \max \left(\sum_{k=1}^{p} (a_{\psi(k)j} - a_{\psi(k)i}), \sum_{k=1}^{p} (a_{\psi(k)i} - a_{\psi(k)j}) \right)$$
$$+ \sum_{(i<j<k) \notin R(p)} \max(|a_{ik} - a_{ij}|, |a_{ik} - a_{jk}|, |a_{ij} - a_{jk}|)$$

122   9 Seriation—Maximization of Gradient Indices

$$f_{Wrc} = \sum_{i=1}^{p-2} \sum_{j=i+1}^{p-1} \sum_{k=j+1}^{p} (2a_{\psi(i)\psi(k)} - a_{\psi(i)\psi(j)} - a_{\psi(j)\psi(k)}) \qquad (9.6.4)$$

$$+ \sum_{i=1}^{p-1} \sum_{j=i+1}^{p} \sum_{k \notin R(p)} (2a_{\psi(i)k} - a_{\psi(i)\psi(j)} - a_{\psi(j)k})$$

$$+ \sum_{(i<j) \notin R(p)} \max\left(\sum_{k=1}^{p} (2a_{\psi(k)j} - a_{\psi(k)i} - a_{ij}), \sum_{k=1}^{p} (2a_{\psi(k)i} - a_{\psi(k)j} - a_{ji}) \right)$$

$$+ \sum_{(i<j<k) \notin R(p)} \max((2a_{ik} - a_{ij} - a_{jk}), (2a_{ij} - a_{ik} - a_{jk}), (2a_{jk} - a_{ik} - a_{ij}))$$

The algorithmic psuedocode for the bound tests for the unweighted cases entails conditional statements to determine the sign function values. For the weighted cases, a function to find maximum values in triplets must be used. Here, we present the algorithmic psuedocode for the unweighted gradient within rows and columns.

/* **Determine** S/R(*Position*) */
Index = *Position* + 1
  for i = 1 to *n*
    found = False
    for j = 1 to *Position*
     if *permutation*(j) = i then found = True
    next j
    if not found then
     complement(Index) = i
     Index = Index + 1
    end if
  next i
/* **Initialize** and **calculate** the upper bound */
UpperB = 0
for i = 1 to *Position* – 2
  for j = i + 1 to *Position* – 1
    for k = j + 1 to *Position*
     difference = **A**(*permutation*(i), *permutation*(k))
                            – **A**(*permutation*(i), *permutation*(j))
     if difference > 0 then UpperB = UpperB + 1
     if difference < 0 then UpperB = UpperB – 1
     difference = **A**(*permutation*(i), *permutation*(k))
                            – **A**(*permutation*(j), *permutation*(k))
     if difference > 0 then UpperB = UpperB + 1

```
   if difference < 0 then UpperB = UpperB – 1
  next k
  next j
next i
for i = 1 to Position – 1
 for j = i + 1 to Position
  for k = Position + 1 to n
   difference = A(permutation(i), complement(k))
                              – A(permutation(i), permutation(j))
   if difference > 0 then UpperB = UpperB + 1
   if difference < 0 then UpperB = UpperB – 1
   difference = A(permutation(i), complement(k))
                              – A(permutation(j), complement(k))
   if difference > 0 then UpperB = UpperB + 1
   if difference < 0 then UpperB = UpperB – 1
  next k
 next j
next i
b = (n – Position) * (n – Position – 1) * (n – Position – 2)
UpperB = UpperB + Position * (n – Position) * (n – Positon – 1) + b/3
```

As with all bound tests, we compare the lower bound of the objective function with the calculated possible upper bound for a partial sequence, UpperB >= LowerB, to determine the feasibility of that partial sequence leading to an optimal solution.

9.3 EXAMPLE—An Archaeological Exploration

Robinson (1951) presents data regarding a Kabah collection in terms of "indexes of agreement" for pairs of types of pottery. A higher index of agreement reflects a closer temporal (chronological) relationship between two deposits. The indexes here record similarity of percentages of pottery types in two deposits. If percentages of pottery types are similar in two deposits, then the deposits were probably made around the same time. Because this is a similarity matrix, we convert to a dissimilarity matrix by subtracting all entries from 200 (see Hubert & Arabie, 1986).

9 Seriation—Maximization of Gradient Indices

Table 9.1. (Dis)Agreement indices for the Kabah collection with rows and columns labeled according to the 17 deposit identifications.

	II	VII	IA	XIA	IB	XB	IX	XA	XIB
II	200	108	68	96	99	116	105	112	106
VII	108	200	95	76	93	84	92	87	94
IA	68	95	200	47	55	65	54	62	50
XIA	96	76	47	200	56	58	50	49	32
IB	99	93	55	56	200	53	34	33	46
XB	116	84	65	58	53	200	36	31	34
IX	105	92	54	50	34	36	200	19	30
XA	112	87	62	49	33	31	19	200	32
XIB	106	94	50	32	46	34	30	32	200
VIII	108	109	72	60	41	43	32	42	47
IVA	119	93	65	66	54	45	33	33	54
VB	128	108	74	79	50	46	36	40	51
VA	145	116	90	93	61	58	52	49	71
VIB	154	118	100	87	53	68	60	57	71
VIA	156	140	109	100	67	71	62	60	72
III	149	122	98	124	69	81	61	57	80
IVB	151	136	95	101	86	69	63	67	79

	VIII	IVA	VB	VA	VIB	VIA	III	IVB	
II	108	119	128	145	154	156	149	151	-
VII	109	93	108	116	118	140	122	136	-
IA	72	65	74	90	100	109	98	95	-
XIA	60	66	79	93	87	100	124	101	-
IB	41	54	50	61	53	67	69	86	-
XB	43	45	46	58	68	71	81	69	-
IX	32	33	36	52	60	62	61	63	-
XA	42	33	40	49	57	60	57	67	-
XIB	47	54	51	71	71	72	80	79	-
VIII	200	53	41	51	52	57	66	61	-
IVA	53	200	47	48	57	73	48	43	-
VB	41	47	200	22	46	36	51	48	-
VA	51	48	22	200	29	28	34	39	-
VIB	52	57	46	29	200	25	48	55	-
VIA	57	73	36	28	25	200	55	61	-
III	66	48	51	34	48	55	200	46	-
IVB	61	43	48	39	55	61	46	200	-

We can permute the rows and columns to optimize the gradient indices, as shown in Table 9.2. For all initial lower bounds, the dynamic ranking permutation of graduated rowsums is $\psi = (1, 2, 3, 4, 17, 16, 15,$

14, 13, 1, 12, 5, 10, 6, 9, 7, 8) with row/column numbers relative to Table 9.1 as are the optimal permutations in Table 9.2, which are then translated to sequential ordering of archaeological deposits. These results show the varied nature of the gradient indices. Naturally, the unweighted gradients have smaller objective function values because they tally the signs of differences in the matrix, whereas the weighted indices accumulate raw differences and become much larger. Very often, when we compare and contrast weighted and unweighted measures, we refer to *sign* and *magnitude* of differences. The weighted and unweighted within row and column gradient indices find optimal permutations that differ by only one change—specifically, objects 7 and 8 (deposits XA and IX) swap places. (As we shall see in Chapter 11, the optimal permutation for the weighted within row and column gradient index is an alternative optimal permutation for the unweighted within row and column gradient index that was ignored by the forward branching algorithm.) However, the optimal permutations for weighted and unweighted within row indices have quite a few differences, indicating the influence that magnitude of differences can have on optimal orderings.

Table 9.2. Optimal seriation for gradient indices of the Kabah collection.

Gradient Index	ψ^*	$f(\psi^*)$
U_r	(16, 17, 13, 14, 12, 15, 11, 10, 8, 7, 5, 6, 9, 4, 3, 2, 1) (III, IVB, VA, VIB, VB, VIA, IVA, VIII, XA, IX, IB, XB, XIB, XIA, IA, VII, II)	562
U_{rc}	(1, 2, 3, 4, 9, 5, 6, 7, 8, 10, 11, 12, 13, 14, 15, 16, 17) (II, VII, IA, XIA, XIB, IB, XB, IX, XA, VIII, IVA, VB, VA, VIB, VIA, III, IVB)	1050
W_r	(17, 16, 15, 14, 13, 12, 11, 10, 7, 8, 6, 5, 9, 4, 3, 2, 1) (IVB, III, VIA, VIB, VA, VB, IVA, VIII, IX, XA, XB, IB, XIB, XIA, IA, VII, II)	23844
W_{rc}	(1, 2, 3, 4, 9, 5, 6, 8, 7, 10, 11, 12, 13, 14, 15, 16, 17) (II, VII, IA, XIA, XIB, IB, XB, XA, IX, VIII, IVA, VB, VA, VIB, VIA, III, IVB)	37980

Another aspect of differences in the optimal permutations for gradient indices is in the difference between within row gradient indices and within row and column gradient indices. The within row gradient indices require optimal permutations to form in a particular direction. In contrast, the within row and column gradient indices are influenced by symmetry in the matrices so that the reverse of an optimal permutation is an alternative optimal permutation. For the above example using forward branching in the branch-and-bound algorithm, the optimal within row and column gradients are found as early as possible, i.e., when object 1 is in the first position rather than when the opposite end of the permutation is in the leading position. The result is that the within row gradients appear to be maximized for the reverse (or near-reverse) optimal permutation for the within row and column gradient indices—yet, that is not the case because the within row and column gradients are optimized with their reverse optimal permutations! Again, as we shall see in Chapter 11, these alternative optimal permutations can be discovered when examining multiple criteria simultaneously. Moreover, in Chapter 10, we will force our main algorithm to take advantage of symmetry in matrices and optimal permutations.

9.4 Strengths and Limitations

The branch-and-bound approach seems to be particularly well suited to the maximization of gradient indices. Brusco (2002b) showed that branch-and-bound was often successful at providing optimal solutions for matrices that were too large for dynamic programming implementation. When enhancing the branch-and-bound approach with insertion and interchange tests, optimal seriation of matrices with 35 to 40 objects is often possible. Although integer programming can also be used for seriation via gradient indices of dissimilarity matrices (Brusco, 2002a), the formulations can be fairly tricky and seem to require more CPU time than branch-and-bound for problems of comparable size.

Gradient indices comparable to those described in this chapter can also be developed for asymmetric proximity matrices (Hubert, 1987; Hubert & Golledge, 1981). However, the optimal permutations based on such criteria can differ violently from those associated with the dominance index. Accordingly, Hubert et al. (2001, Chapter 4) recommend avoidance of the gradient indices for asymmetric matrices in favor of the dominance index.

9.5. Available Software

Located at websites http://www.psiheart.net/quantpsych/mongraph.html and http://garnet.acns.fsu.edu/~mbrusco, four software programs are available for finding a permutation of the rows and columns of a symmetric dissimilarity matrix so as to maximize a gradient index, which are much-improved versions of programs originally described by Brusco (2002b). These programs are *bburg.for* (a program that maximizes the unweighted within-row gradient index), *bbwrg.for* (a program that maximizes the weighted within-row gradient index), *bburcg.for* (a program that maximizes the unweighted within-row and within-column gradient index), and *bbwrcg.for* (a program that maximizes the weighted within-row and within-column gradient index).

All four gradient index programs use the same file structure. The first line of the input data file, *sym.dat*, must contain the number of objects, n, and succeeding lines should contain the dissimilarity data. Like many of the programs described in previous chapters, the user will be prompted for the form of the matrix, and should enter 1 for a half matrix or 2 for a full $n \times n$ matrix. The output of the program is written both to the screen and to a file "*results*." The output includes the optimal permutation of objects, the lower bound for the index value provided by a heuristic, the optimal index value, the CPU time required to obtain the solution, and the optimal permutation.

Like *bbdom.for*, each of the four gradient-index programs uses a pairwise interchange heuristic to establish a lower bound for the index. The main algorithm is then initiated and employs pairwise interchange tests, single-object relocation tests, and bound tests to prune solutions. As described in section 8.6, the single-object relocation tests are similar to interchange tests.

To illustrate the four gradient index programs, we present the results corresponding to the application of each program to a 30×30 food-item dissimilarity matrix published by Hubert et al. (2001, pp. 100-102) and based on data originally collected by Ross and Murphy (1999). More specifically, these data correspond to the first 30 rows and columns of the larger 45×45 dissimilarity matrix published by Hubert et al. (2001). The 30×30 matrix was also analyzed by Brusco (2002b). Below is the output for each program:

Output for *bburg.for*:
HEURISTIC OBJ VALUE   2986
MAXIMUM UNWEIGHTED ROW GRADIENT INDEX 3093

CPU TIME 103.09
1 2 3 4 5 6 7 8 9 10 11 12 24 19 13 14 16 15 20 17 18 21 22 23 25 26 27 28 29 30

Output for *bburcg.for*:
OBJ VALUE   5580
MAXIMUM UNWEIGHTED ROW AND COLUMN GRADIENT INDEX 5580
CPU TIME 25.62
1 2 3 4 5 6 7 9 8 10 11 12 24 19 13 14 15 16 20 17 18 21 22 23 25 26 27 28 29 30

Output for *bbwrg.for*:
HEURISTIC OBJ VALUE   91085.0000
MAXIMUM WEIGHTED ROW GRADIENT INDEX   91085.0000
CPU TIME 64.00
1 2 3 4 5 6 7 9 8 10 11 12 19 13 14 16 15 18 17 20 21 22 24 23 25 26 27 28 29 30

Output for *bbwrcg.for*:
HEURISTIC OBJ VALUE   167279.0000
MAXIMUM WITHIN ROW AND COLUMN GRADIENT INDEX 167279.0000
CPU TIME 4.41
1 2 3 4 5 6 7 9 8 10 11 12 19 13 14 16 15 18 17 20 21 22 24 23 25 26 27 28 29 30

A couple of comments regarding the results of the gradient index programs are in order. First, the permutations across the four indices are remarkably similar. (In fact, the optimal permutations for the two weighted gradient indices are identical.) Other dissimilarity matrices can produce markedly different results across the four indices.

Second, the computation times for these 30×30 matrices are rather modest and compare very favorably to those reported by Brusco (2002b). Part of the reduction in CPU time is certainly attributable to the fact that the 2.2 GHz Pentium IV that we used to obtain the results in this monograph is appreciably faster than the 666 MHz Pentium III used by Brusco (2002b). However, most of the improvement is attributable to the inclusion of better bounding procedures and more efficient computational schemes in the newer programs. Most notably, the inclusion of interchange and single-object relocation tests to prune solutions, which were not considered by Brusco (2002b), produce substantial reduction in CPU time.

10 Seriation—Unidimensional Scaling

10.1 Introduction to Unidimensional Scaling

Thus far, we have discussed seriation to optimize patterning in a matrix according to simple mathematical comparisons ("greater than"/"less than") of matrix entries. However, with available data, we often need to know how "close" the objects are. In particular, we would like a graphical representation on a number line to show us the relationship between the objects. At this juncture, we delve into unidimensional scaling to help us better observe the nature of the data.

If we assign objects from a symmetric dissimilarity matrix, **A**, to actual coordinates on a number line, then we can measure how well the coordinates graphically represent the data using the least-squares loss function. To find the best, if not perfect, coordinates to represent the data, we minimize the least-squares loss function as in equation (10.1):

$$Min: Z = \sum_{i<j}(a_{ij} - |x_i - x_j|)^2 . \qquad (10.1)$$

For a symmetric dissimilarity matrix, **A**, the minimization of the least-squares loss function is tantamount to the combinatorial maximization problem formulated by Defays (1978):

$$Max: g(\psi) = \sum_{i=1}^{n}(\sum_{j=i+1}^{n} a_{\psi(i)\psi(j)} - \sum_{j=1}^{k-1} a_{\psi(i)\psi(j)})^2 \text{ for } \psi \in \Psi . \qquad (10.2)$$

An important note to make is that anti-Robinson structure is indicative of scalability on a number line. In other words, the closer a matrix adheres to anti-Robinson patterning, the smaller the error tends to be when scaling the objects. In fact, "error-free" unidimensional scaling implies complete anti-Robinson structure; however, the converse is not necessarily true. Of course, other characteristics are also desirable in the data prior to scaling.

Using Defays' maximization, the algorithmic pseudocode that should be implemented to find a permutation to optimally scale an $n \times n$ symmetric dissimilarity matrix, **A**, is as follows:

```
UDS = 0
for i = 1 to n
   summation = 0
   for j = 1 to n
      if i > j then summation = summation
                                 + A(permutation(i), permutation(j))
      if i < j then summation = summation
                                 - A(permutation(i), permutation(j))
   next j
   UDS = UDS + summation * summation
next i
```

We use this EVALUATION in our main branch-and-bound algorithm to find a permutation of the rows and columns of the matrix **A** that will produce the maximum UDS variable. Once the permutation has been found, we can find the appropriate coordinates for the n objects:

```
for i = 1 to n
   coordinate(i) = 0
   for j = 1 to n
      if i < j then coordinate(i) = coordinate(i)
                                      - A(permutation(i), permutation(j))
      if i > j then coordinate(i) = coordinate(i)
                                      + A(permutation(i), permutation(j))
   next j
   coordinate(i) = coordinate(i)/n
next i
```

The coordinates are centered about 0 but can be shifted to begin at 0 by subtracting the first coordinate from all n coordinates. The error is calculated according to the least-squares loss function:

```
CError = 0
for i = 1 to n
   for j = 1 to n
      difference = (A(permutation(i), permutation(j))
                      - abs(coordinate(i) - coordinate(j)))
      CError = CError + difference * difference
   next j
next i
```

10.2 Fathoming Tests for Optimal Unidimensional Scaling

10.2.1 Determining an Initial Lower Bound

Unidimensional scaling tends to put objects with higher entries at opposite ends of the sequence. Therefore, a prudent initial lower bound would be the evaluation of a permutation for which objects with higher *rowsums* (with a zero diagonal) alternate at the ends of the sequence, working toward the middle of the sequence. Fortunately, we can use the highest-to-lowest ranking function of equations (8.3) and (8.3.1), or the ranking of the graduated *rowsums* as described in section 8.2.1. This is then followed by a reordering of the ranks to mimic the objective function for unidimensional scaling as in the *ends_rank* function of equation (9.3). Finally, as we did to initialize the main algorithm for maximizing gradient indices, we use a pairwise interchange heuristic to quickly approach a local optimum.

10.2.2 Testing for Symmetry in Optimal Unidimensional Scaling

One of the characteristics of unidimensionally scaling a symmetric matrix is that the reverse order of an optimal permutation yields another optimal permutation. This is intuitively apparent because two objects on a number line are the same distance apart regardless of which object is placed first on the number line. For example, if two objects are placed at points 2 and 6, then they are separated by four units regardless of which object is placed at 2 or 6. We can use this property to our advantage during the implementation of the branch-and-bound algorithm.

With two integral values, we could arbitrarily choose to ensure that one object precedes another in the optimal permutation, pruning all sequences for which the second value is assigned a position prior to assigning the first value. However, we can improve our strategy by wisely choosing two global integral values. Because we know the nature of the objective function and we have determined a ranking order of *rowsums* for our n objects, we can choose the object values that will tend to be placed far apart in the sequence. When determining the initial lower bound for the objective function value, we can select two values to check for symmetry in our branch-and-bound algorithm, *SymFirst = permuta-*

tion(1) and *SymSecond* = *permutation*(2), where the permutation has been dynamically ranked and used to evaluate our initial lower bound. More than that, with an understanding of the process by which the branch-and-bound algorithm generates permutations for evaluation, we can choose the order of the symmetry. Once the permutation has the highest-to lowest ordering, we check objects in the permutation relative to their values with a simple conditional statement.

if *permutation*(1) < *permutation* (2) then
   SymFirst = *permutation*(1) and *SymSecond* = *permutation*(2)
else
   SymFirst = *permutation*(2) and *SymSecond* = *permutation*(1)
end if

With these variables in place, we are poised to check for our symmetry condition in the main algorithm immediately following the redundancy check. Although actual implementation may vary by using Boolean or binary triggers, a brute force check after the redundancy check is straightforward and fairly efficient.

NotSymmetry = True
if (*NotRedundancy* and *permutation*(*Position*) = *SymSecond*) then
   NotSymmetry = False
   for i = 1 to *Position* − 1
     if *permutation*(i) = *SymFirst* then *NotSymmetry* = True
   next i
end if

If the symmetry condition is not found (i.e., *NotSymmetry* remains True), then we may continue to build our permutation toward a possible optimal solution. However, if the symmetry flag (*SymSecond*) is found and the *NotSymmetry* condition is not found during the loop of the conditional statement, then we ignore any further evaluation or fathom testing until we change the object value in the current *Position* of the *permutation* essentially branching right. Hence, we would only proceed with the main algorithm if we have both "not redundancy" and "not symmetry" conditions met. This symmetry check can prune many branches and conserve computation time significantly.

10.2.3 Adjacency Test Expanded to Interchange Test

For finding an optimal order by which to unidimensionally scale the n objects corresponding to rows and columns of a symmetric dissimilarity

10.2 Fathoming Tests for Optimal Unidimensional Scaling

matrix, the branch-and-bound method uses an interchange test in lieu of an adjacency test. The interchange test finds the maximum contribution to the objective function—or, equivalently, the minimal contribution to the least-squares loss function—by swapping the object in *Position* with the objects in all other previously assigned positions. As such, the interchange test is more computationally intensive than simply swapping the objects in *Position* and *Position* − 1. However, the computational effort is more than compensated by the pruning of many more branches that would otherwise be needlessly fathomed (Brusco & Stahl, 2004).

To begin understanding the effect of swapping objects in two positions of a sequence, p and q (WOLOG $p > q$), consider how the objective function value is computed. For each row, the objective function for unidimensional scaling sums the entries on the left of the diagonal, sums the entries on the right of the diagonal, and squares the difference between the sums. Therefore, the objective function is only affected by entries that "switch sides" of the diagonal. Figure 10.1 shows the entries that are affected by the change.

		Position q			Position p		
	0	(x)	x	x	(x)	x	x
Position q	(x)	0	switch	switch	switch	(x)	(x)
	x	switch	0	x	switch	x	x
	x	switch	x	0	switch	x	x
Position p	(x)	switch	switch	switch	0	(x)	(x)
	x	(x)	x	x	(x)	0	x
	x	(x)	x	x	(x)	x	0

Figure 10.1. An illustration of the change to the placement of matrix entries after two objects swap positions. The diagonal is zeroed, "x" indicates no change in placement, "(x)" indicates a change in placement that will occur on the *same side* of the diagonal, and "switch" indicates an entry that will "switch sides" of the diagonal.

At this stage, we are only observing the changes in the placement of entries when two objects swap positions. However, we note that the interchange test will be performed when only the first p positions of the permutation are filled. As Figure 10.1 shows, this is not a problem because the entries affected in rows following p will not switch sides of the diagonal—lower to upper triangle or vice versa—as indicated by x's in parentheses in rows/columns following row p.

For each row (object) i of the matrix, entries to the left of the diagonal are summed, $\text{SumLeft}(i) = \sum_{j=1}^{i-1} a_{ij}$. As with the methodology for maximizing the dominance index, we can conveniently calculate the *rowsums*, r_i, ignoring the diagonal entries, for each object prior to implementing the branch-and-bound algorithm so that $\text{SumRight}(i) = \sum_{j=i+1}^{n} a_{ij} = r_i - \text{SumLeft}(i)$. The objective function sums $(\text{SumLeft}(i) - \text{SumRight}(i))^2 = (\text{SumLeft}(i) - (r_i - \text{SumLeft}(i)))^2 = (2*\text{SumLeft}(i) - r_i)^2$. Therefore, for a given permutation, ψ, the contribution of rows q through p to the objective function is

$$\sum_{i=q}^{p}(2*\text{SumLeft}(\psi(i))-r_{\psi(i)})^2 = \sum_{i=q}^{p}(2*\sum_{j=1}^{i-1} a_{\psi(i)\psi(j)} - r_{\psi(i)})^2. \quad (10.3)$$

Because the matrix is symmetric, the adjacency test simplifies greatly because the only change will be the addition of $a_{\psi(p-1)\psi(p)}$ to $\text{SumLeft}(\psi(p-1))$ and the subtraction of $a_{\psi(p)\psi(p-1)}$ from $\text{SumLeft}(p)$. However, for the more comprehensive interchange test, we consider swapping the sequence positions of objects currently in positions p and q. For rows prior to q and subsequent to p, matrix entries a_{iq} and a_{ip} exchange places but do not cross the diagonal, having no effect on the $\text{SumLeft}(i)$ calculation. However, for rows between q and p ($q < i < p$), a_{iq} and a_{ip} exchange places and "switch sides" of the diagonal in row i when p and q are swapped. Hence, if we denote the reordered row of i as $\psi'(i)$, then $\text{SumLeft}(\psi'(i)) = \text{SumLeft}(\psi(i)) + a_{\psi(i)\psi(p)} - a_{\psi(i)\psi(q)}$ for $q < i < p$. For the rows being swapped, p and q, every entry in rows p and q between columns p and q will either join or leave the $\text{SumLeft}(i)$ for that row. That is, $\text{SumLeft}(\psi'(q)) = \text{SumLeft}(\psi(q)) + \sum_{i=q+1}^{p} a_{\psi(q)\psi(i)}$ and $\text{SumLeft}(\psi'(p)) = \text{SumLeft}(\psi(p)) - \sum_{i=q}^{p-1} a_{\psi(p)\psi(i)}$. In short, only the entries in rows p and q and columns p and q will actually affect the objective function value for unidimensional scaling. Because the *rowsums*

10.2 Fathoming Tests for Optimal Unidimensional Scaling

remain constant, we can state, expand, and simplify the contribution to the objective function by rows p through q after exchanging the sequence positions of p and q as follows:

$$\sum_{i=q}^{p} (\text{SumLeft}(\psi'(i)) - \text{SumRight}(\psi'(i)))^2$$

$$= \sum_{i=q}^{p} (2 * \text{SumLeft}(\psi'(i)) - r_{\psi'(i)})^2$$

$$= (\text{SumLeft}(\psi'(q)) - r_{\psi'(q)})^2 + (\text{SumLeft}(\psi'(p)) - r_{\psi'(p)})^2 + \sum_{i=q+1}^{p-1} (2 * \text{SumLeft}(\psi'(i)) - r_{\psi'(i)})^2$$

$$= ((\text{SumLeft}(\psi(q)) + \sum_{i=q+1}^{p} a_{\psi(i)\psi(q)}) - r_{\psi(q)})^2 + ((\text{SumLeft}(\psi(p)) - \sum_{i=q}^{p-1} a_{\psi(i)\psi(p)}) - r_{\psi(p)})^2$$

$$+ \sum_{i=q+1}^{p-1} (2 * (\text{SumLeft}(\psi(i)) - a_{\psi(i)q} + a_{\psi(i)p}) - r_{\psi(i)})^2$$

$$= (\sum_{j=1}^{q-1} a_{\psi(q)\psi(j)} + \sum_{i=q+1}^{p} a_{\psi(i)\psi(q)} - r_{\psi(q)})^2 \quad (10.4)$$

$$+ (\sum_{j=1}^{p-1} a_{\psi(p)\psi(j)} - \sum a_{\psi(i)\psi(p)} - r_{\psi(p)})^2$$

$$+ \sum_{i=q+1}^{p-1} (2 * (\sum_{j=1}^{i-1} a_{\psi(i)\psi(j)} - a_{\psi(i)\psi(q)} + a_{\psi(i)\psi(p)}) - r_{\psi(i)})^2$$

Although equation (10.4) might seem arduous, the repetition of similar terms lends itself to ease of implementation. The pseudocode for calculating and comparing the contributions of rows p through q before and after the exchange is modular.

Set AltPosition = q and *Position* = p.
CurrentContribution = 0
AltContribution = 0
for i = AltPosition to *Position*
    /* **Calculate** the contribution to the objective function by row *i* prior to exchanging the positions of objects located in positions p and q. */
    SumLeft = 0
    for j = 1 To i - 1
        SumLeft = SumLeft + **A**(*permutation*(i), *permutation*(j))
    next j
    Contribution = 2 * SumLeft − *rowsum*(*permutation*(i))
    CurrentContribution = CurrentContribution
                               + Contribution * Contribution

/* **Calculate** the contribution to the objective function by row i after exchanging the positions of objects located in positions p and q. */
```
if (i > AltPosition And i < Position) then
    SumLeft = SumLeft
                    - A(permutation(i), permutation(AltPosition))
                    + A(permutation(i), permutation(Position))
end if
if i = AltPosition then
    for j = AltPosition + 1 to Position
        SumLeft = SumLeft
                        + A(permutation(AltPosition), permutation(j))
    next j
end if
if i = Position then
    for j = AltPosition to Position - 1
        SumLeft = SumLeft
                        - A(permutation(Position), permutation(j))
    next j
end if
Contribution = 2 * SumLeft - rowsum(permutation(i))
AltContribution = AltContribution + Contribution * Contribution
next i
```

For the complete interchange test, we perform this exchange test between rows p and $q = 1,\ldots, p-1$, i.e., between *Position* and *AltPosition* $= 1,\ldots,$ *Position* $- 1$. If the CurrentContribution is less than the AltContribution for any object AltPosition in positions 1 through *Position* $- 1$, then the INTERCHANGETEST fails, and we increment the object in Position and continue.

10.2.4 Bound Test

Given a partial sequence of p out of n objects, the upper bound for the objective function value for unidimensional scaling is comprised of two components. The first component calculates the contribution of the first p objects to the objective function value:

$$f_{U1} = \sum_{i=1}^{p}(r_{\psi(i)} - 2\sum_{j=1}^{i-1} a_{\psi(i)\psi(j)})^2 . \qquad (10.5)$$

```
UpperB1 = 0
for i = 1 to Position
    temp = 0
    for j = 1 to i – 1
        temp = temp + A(permutation(i), permutation(j))
    next j
    temp = rowsum(permutation(i)) – 2*temp
    temp = temp*temp
    UpperB1 = UpperB1 + temp
next i
```

The second component is an upper bound on the possible contribution by the remaining $n - p$ objects. For unselected objects in unfilled positions, the upper bound should find the maximum possible contribution to the objective function of object i if placed in position j. For any object, the set of row entries is divided into two groups—those to the left of the diagonal and those to the right of the diagonal. For unidimensional scaling, the goal is to have the largest possible difference between the sums of the two groups. Therefore, we should put the smaller entries into the smaller of the subsets and the larger entries in the larger of the subsets. That is, if object i is placed in position p and $p \leq n / 2$, then the most desirable outcome is to have the smallest entries to the left of the diagonal; similarly, if object i is placed in position p and $p > n / 2$, then the most desirable outcome is to have the smallest entries to the right of the diagonal. So, for a row i, we find the pth greatest row entry with,

$$order(i, p) = [a_{ij} \mid p = \{a_{ik} : a_{ij} < a_{ik}\} \mid +1]. \tag{10.6}$$

Now, with this helpful concept, we know that if $p > n / 2$, then: $\text{BigSum}(i, p) = \sum_{j=1}^{p-1} order(i, j)$. Moreover, based on the SumLeft and SumRight functions, $\text{SmallSum}(i, p) = rowsum(i) - \text{BigSum}(i, p)$.

Rather than repeatedly calculating these values during execution of the bound test, we can calculate and record these values for all n objects in all n positions prior to implementation of the main algorithm, i.e. immediately following calculation of the initial lower bound. Once the lowest-to-highest ordering of entries for each row has been made, a matrix **B** can be constructed to hold the most desirable outcome for each object in each position. The second component can then be computed quickly for every bound test performed during execution by finding the maximum "most desirable" contribution of the unselected objects for each unfilled position. We establish the **B** matrix:

for object = 1 to n
  /* **Order** entries in row *object* from lowest to highest. The following conditional loop ignores the diagonal entry of **A**(object, object) = 0 when initializing the *Order* array */
  for Index = 1 to n
    if Index < object then Order(Index) = **A**(object, Index)
    if Index > object then Order(Index − 1) = **A**(object, Index)
  next Index
  for i = 1 to n − 1
    Minimum = Order(i)
    for j = i + 1 fo n − 1
      if Order(j) < Minimum then
        Minimum = Order(j), Order(j) = Order(i), Order(i) = Minimum
      end if
    next j
  next i
  /* **Calculate** b_{ij} for all positions of object in permutation */
  for *Position* = 1 to n
    SumLeft = 0
  if *Position* < ((n / 2) + 1) then
    for i = 1 to *Position* − 1
      SumLeft = SumLeft + Order(i)
    next i
    SumRight = *rowsum*(object) − SumLeft
    B(object, *Position*) = (SumRight − SumLeft)*(SumRight − SumLeft)
  else
    B(object, *Position*) = **B**(object, n − *Position* + 1)
  end if
  next *Position*
next object

Now, in the bound test, when we want to find the maximum possible contribution of an unselected object i in a particular position j, we simply recall b_{ij}. In determining the second component of the upper bound, we only consider objects in $S \setminus R(p)$.

UpperB2 = 0
/* **Determine** $S \setminus R$(*Position*). */
index = *Position* + 1
for i = 1 to n

```
    found = False
    for j = 1 to Position
        if permutation(j) = i then found = true
    next j
    if not found then
        complement(index) = i
        index = index + 1
    end if
next i
/* Find maximum contribution to objective function value by the
complement. */
for OpenPosition = Position + 1 to n
    for Object = Position + 1 to n
        if B(complement(Object), OpenPosition) > MAX then
        MAX = B(complement(Object), OpenPosition)
        end if
    next Object
    UpperB2 = UpperB2 + MAX
next OpenPosition
```

10.3 Demonstrating the Iterative Process

Consider the sample symmetric dissimilarity matrix in Table 10.1. To begin unidimensional scaling, we find the permutation of graduated *rowsums* of $\psi = (4, 2, 1, 3)$. After alternating the *rowsums* at the ends of the permutation toward the middle and then approaching a local optimum, we have $\psi = (2, 1, 3, 4)$ with *SymFirst* = 1, *SymSecond* = 2 and an initial lower bound of $f(\psi) = 6316$.

Table 10.1. A sample symmetric dissimilarity matrix.

	1	2	3	4
1	0	15	9	25
2	15	0	12	20
3	9	12	0	16
4	25	20	16	0

With the initial lower bound, the iterative process in the main algorithm is begun as shown in Table 10.2. During the iterative process, the symmetry check and interchange test pruned quite a few branches. The bound test was also a useful pruning tool, as on row 18. The incumbent solution was improved twice, finally settling on the optimal solution of 6532 on row 12 with the permutation $\psi = (1, 3, 2, 4)$. Whenever the algorithm attempted to add a fifth element to the object list, retraction occurred to begin a new branch. Once the permutation was known, the coordinates were calculated at [0, 7.5, 14, 27.5] with an error of 98.

Table 10.2. The iterative process for unidimensional scaling of the symmetric dissimilarity matrix in Table 10.1.

Row	Partial Sequence	Interchange test	UpperB1	UpperB2	UpperB	Dispensation
1	1 - 2	Pass	2690	4562	7252	Branch, (7252 > 6316)
2	1 - 2 - 3					New incumbent, $f^* = 6436$
3	1 - 2 - 4					Suboptimal, (4900 < 6436)
4	1 - 2 - 5					Retraction
5	1 - 3	Pass	2762	4562	7324	Branch, (7324 > 6436)
6	1 - 3 - 2					New incumbent, $f^* = 6532$
7	1 - 3 - 4					Suboptimal, (5412 < 6532)
8	1 - 3 - 5					Retraction
9	1 - 4	Fail				Prune, Exchange test failed for objects 4 and 1
10	1 - 5					Retraction
11	2					Fail Symmetry
12	3	Pass	1369	5643	7012	Branch, (7012 > 6532)
13	3 - 1	Fail				Prune, Exchange test failed for objects 1 and 3
14	3 - 2					Fail Symmetry
15	3 - 4	Fail				Prune, Exchange test failed for objects 4 and 3
16	3 - 5					Retraction
17	4	Pass	3721	4323	8044	Branch, (8044 > 6532)
18	4 - 1	Pass	3722	2738	6460	Prune, (6460 < 6532)
19	4 - 2					Fail Symmetry
20	4 - 3	Pass	3746	3362	7108	Branch, (7108 > 6532)
21	4 - 3 - 1					Suboptimal, (6316 < 6532)
22	4 - 3 - 2					Fail Symmetry
23	4 - 3 - 5					Retraction
24	4 - 5					Retraction
25	5					TERMINATE

10.4 EXAMPLE—Can You Hear Me Now?

As an example of a branch-and-bound application to empirical proximity data, consider the information in Table 10.3.

Table 10.3. Confusion data for English consonants degraded by –18db (Miller & Nicely, 1955).

	(p)	(t)	(k)	(f)	(θ)	(s)	(ʃ)	(b)
(p)	.000	.102	.083	.087	.095	.083	.053	.057
(t)	.073	.000	.095	.068	.068	.082	.064	.032
(k)	.083	.092	.000	.063	.058	.121	.050	.017
(f)	.101	.082	.101	.000	.049	.045	.037	.071
(θ)	.071	.075	.075	.054	.000	.088	.050	.058
(s)	.071	.067	.091	.044	.071	.000	.067	.044
(ʃ)	.060	.075	.101	.063	.049	.138	.000	.037
(b)	.045	.041	.090	.056	.071	.056	.045	.000
(d)	.054	.081	.061	.044	.051	.051	.047	.074
(g)	.036	.066	.095	.030	.059	.059	.049	.086
(v)	.040	.076	.080	.049	.031	.054	.040	.112
(δ)	.074	.051	.046	.032	.028	.065	.046	.093
(z)	.074	.074	.061	.037	.053	.078	.029	.090
(ʒ)	.036	.073	.077	.064	.055	.068	.032	.100
(m)	.079	.100	.063	.058	.058	.058	.033	.058
(n)	.047	.076	.085	.025	.038	.076	.038	.059

	(d)	(g)	(v)	(δ)	(z)	(ʒ)	(m)	(n)
(p)	.061	.027	.064	.042	.045	.042	.061	.045
(t)	.045	.027	.077	.041	.059	.050	.041	.059
(k)	.046	.038	.050	.042	.067	.046	.071	.058
(f)	.075	.052	.060	.060	.056	.011	.049	.067
(θ)	.083	.058	.096	.025	.058	.038	.050	.058
(s)	.095	.060	.060	.063	.044	.052	.067	.020
(ʃ)	.078	.026	.075	.067	.034	.030	.060	.056
(b)	.075	.071	.090	.045	.056	.041	.067	.063
(d)	.000	.071	.084	.057	.061	.044	.051	.084
(g)	.099	.000	.059	.046	.053	.066	.079	.072
(v)	.063	.058	.000	.067	.085	.049	.054	.076
(δ)	.079	.083	.069	.000	.079	.056	.083	.083
(z)	.057	.037	.086	.049	.000	.041	.090	.049
(ʒ)	.082	.036	.068	.050	.068	.000	.082	.059
(m)	.063	.050	.054	.033	.046	.025	.000	.117
(n)	.059	.055	.038	.034	.042	.051	.140	.000

To prepare the data, we must induce symmetry and dissimilarity. Symmetry is induced by using weighted averages for "mirror" entries as in equation (10.7). With all entries strictly between 0 and 1, we induce dissimilarity for this demonstration by simply subtracting each matrix entry from 1.

$$[a'_{ij}] = \left[\frac{1}{2}\left(\frac{a_{ij}}{r_i} + \frac{a_{ji}}{r_j}\right)\right]. \tag{10.7}$$

The branch-and-bound algorithm finds the optimal permutation of ψ = (3, ð, g, b, v, z, d, n, m, t, k, s, θ, p, f, ʃ) corresponding to coordinates [0, .1208, .2386, .3576, .475, .5886, .707, .8289, .9386, 1.0675, 1.1811, 1.2934, 1.4119, 1.5286, 1.6461, 1.7676] with an error of 29.4549 and the Defays' maximization of $g(\psi^*)$ = 1201.8721. An interesting note is that the closely related matrix of Table 10.4 produces very different results—specifically, ψ = (t, k, p, f, θ, s, ʃ, 3, z, d, g, ð, v, b, m, n), coordinates [0, .0945, .1837, .3278, .4294, .555, .6543, .8488, .9522, 1.0707, 1.1756, 1.306, 1.4164, 1.5145, 1.6968, 1.7482] with an error of 25.9953 and $g(\psi^*)$ = 1270.4331.

Table 10.4. Confusion data for English consonants degraded by –12db (Miller & Nicely, 1955).

	(p)	(t)	(k)	(f)	(θ)	(s)	(ʃ)	(b)
(p)	.000	.207	.254	.086	.074	.023	.043	.008
(t)	.219	.000	.253	.068	.082	.075	.048	.007
(k)	.212	.178	.000	.093	.076	.068	.047	.017
(f)	.121	.086	.109	.000	.113	.059	.043	.012
(θ)	.096	.081	.092	.232	.000	.099	.044	.022
(s)	.069	.065	.069	.142	.103	.000	.207	.013
(ʃ)	.127	.177	.111	.078	.150	.139	.000	.006
(b)	.016	.008	.008	.070	.027	.027	.004	.000
(d)	.013	.000	.004	.017	.030	.017	.047	.078
(g)	.013	.004	.004	.004	.017	.021	.029	.083
(v)	.000	.004	.004	.051	.021	.017	.021	.157
(ð)	.000	.004	.015	.063	.007	.011	.007	.198
(z)	.025	.004	.008	.008	.025	.059	.034	.097
(3)	.013	.009	.009	.004	.000	.026	.030	.030
(m)	.000	.009	.000	.000	.009	.009	.000	.097
(n)	.008	.000	.000	.008	.000	.008	.000	.015

10.4 EXAMPLE—Can You Hear Me Now?

Table 10.4 -Continued

	(d)	(g)	(v)	(ð)	(z)	(ʒ)	(m)	(n)
(p)	.000	.008	.012	.012	.004	.020	.031	.020
(t)	.010	.003	.003	.007	.003	.003	.017	.003
(k)	.017	.004	.004	.008	.000	.000	.017	.008
(f)	.020	.000	.031	.031	.012	.000	.012	.000
(θ)	.033	.011	.040	.033	.011	.007	.026	.007
(s)	.022	.026	.013	.004	.026	.009	.000	.004
(ʃ)	.022	.028	.017	.000	.033	.017	.022	.011
(b)	.070	.070	.172	.098	.055	.023	.078	.039
(d)	.000	.151	.069	.103	.112	.060	.039	.052
(g)	.158	.000	.067	.121	.121	.158	.042	.038
(v)	.085	.097	.000	.068	.059	.017	.059	.038
(ð)	.116	.093	.187	.000	.086	.019	.049	.022
(z)	.123	.114	.102	.081	.000	.110	.013	.025
(ʒ)	.129	.099	.039	.030	.168	.000	.022	.060
(m)	.026	.053	.070	.097	.000	.009	.000	.527
(n)	.015	.046	.054	.008	.008	.070	.649	.000

Unidimensional scaling offers the convenience of graphical displays of seriation, as shown in Figures 10.2 and 10.3. We can easily compare the two graphical representations of the data in terms of spacing as well as ordering. In particular, the more confusing data (−18db) produce more uniform confusion, i.e. more evenly spaced objects. The less confusing data (−12db) place more similar objects closer on the continuum. For example, the consonants *m* and *n*, which are indeed very similar and easily confused, are adjacent to one another in the seriation of either matrix; however, the graphical representation of −12db degradation clearly shows a closer relationship of these consonants to one another with respect to the other consonants. One way to more fully examine the relationship of the two data sets is via multiobjective seriation, which is described in Chapter 11.

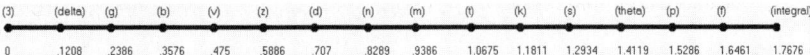

Fig. 10.2 Graphical representation of data in Table 10.3 (−18db degradation).

Fig. 10.3 Graphical representation of data in Table 10.4 (−12db degradation).

10.5 Strengths and Limitations

Unidimensional scaling has an important place in the combinatorial data analysis literature, and Defays' branch-and-bound approach was one of the first methods designed to provide guaranteed optimal solutions for this problem. Brusco and Stahl (2004) have recently extended this approach to increase computational efficiency and facilitate the solution of larger problems. Nevertheless, dynamic programming still tends to be much faster than branch-and-bound when computer memory is sufficient to enable its deployment. Furthermore, we should also recognize, for problems that are computationally feasible for these optimal solution approaches, most of the available heuristic approaches (Groenen, Heiser, & Meulman, 1999; Hubert, Arabie, & Meulman 2002) also tend to produce optimal solutions in much less time.

There are alternatives to the least-squares loss function for unidimensional scaling. Most notable among these is the least-absolute-deviation (or L_1-norm) loss function:

$$Min : Z = \sum_{i<j} | \; a_{ij} - | x_i - x_j | \; |. \tag{10.8}$$

Integer programming formulations for (10.8) have been developed by Simantiraki (1996) and Chen (2000). Brusco (2002a) has noted that neither of these formulations is especially sharp, and they are limited to problems of size $15 \leq n \leq 20$. A reformulation of (10.8) similar to the one provided by Defays (1978) remains undiscovered, and thus effective dynamic programming and branch-and-bound approaches are not currently available. However, Hubert et al. (2002) have recently observed that there seems to be no pragmatic advantage to the least-absolute-deviation criterion over the least-squares criterion, concluding that the more widely accepted least-squares criterion is the measure of choice.

10.6 Available Software

We have made available two software programs for optimal least-squares unidimensional scaling of a symmetric dissimilarity matrix. The first program, *bbinward.for*, uses Defays' (1978) branch-and-bound solution strategy in conjunction with several enhancements described by Brusco and Stahl (2004). These enhancements include the use of an interchange test in addition to the adjacency test, as well some improved bounding procedures. The key aspect of *bbinward.for* is that it assigns objects to positions from the ends of the sequence inward, rather than from left-to-right (forward branching) like all other branch-and-bound seriation approaches described in this monograph. The second program, *bbforwrd.for*, is similar in design to *bbinward.for* except that forward branching is used. These programs, which are slightly improved versions of programs originally described by Brusco and Stahl (2004), can be downloaded from http://www.psiheart.net/quantpsych/monograph.html or http://garnet.acns.fsu.edu/~mbrusco.

Both programs use a pairwise interchange heuristic to establish a lower bound on the Defays criterion. Once this bound is established the branch-and-bound process is initiated using the appropriate tests for pruning partial solutions. Upon completion of the branch-and-bound process, the optimal permutation is used to establish the coordinates that minimize the least-squares loss function. Both unidimensional scaling programs use file structures that are similar to those of the gradient indices. The user is prompted regarding the format of the data in the file *amat.dat* (i.e., 1 for half matrix or 2 for full matrix). The output file contains the value of the heuristic solution for the Defays criterion, the optimal Defays criterion value, the optimal loss function value, the total CPU time required to obtain the optimal solution, the optimal permutation of objects, and corresponding coordinates.

The branch-and-bound programs for least-squares unidimensional scaling are illustrated using the lipread consonant dissimilarity matrix from Table 3.3.

The solution summary output for *bbinward.for* is:
HEURISTIC DEFAYS CRIT VALUE   255912968.0000
MAXIMUM DEFAYS CRITERION VALUE   255912968.00000
CPU TIME   511.56
MINIMUM LOSS FUNCTION VALUE   3082638.19048

The solution summary output for *bforwrd.for* is:
HEURISTIC DEFAYS CRIT VALUE   255912968.0000
MAXIMUM DEFAYS CRITERION VALUE   255912968.00000
CPU TIME   13.19
MINIMUM LOSS FUNCTION VALUE   3082638.19048

Both programs produce the same optimal permutation and coordinates. The three columns of output below are the position number, object index placed in that position, and object coordinate, respectively.

1	10	0.00000
2	4	31.38095
3	14	62.14286
4	9	101.71429
5	15	128.09524
6	19	135.66667
7	6	167.57143
8	11	187.19048
9	8	214.80952
10	12	278.23810
11	1	288.09524
12	21	319.42857
13	2	325.23810
14	3	338.80952
15	16	350.76190
16	17	388.71429
17	5	408.52381
18	7	449.33333
19	13	496.33333
20	20	552.80952
21	18	598.14286

One of the most important aspects of the results for the lipread consonant data is the disparity between the CPU times for *bbinward.for* and *bbforwrd.for*, with the former requiring nearly 40 times more CPU time than the latter. Brusco and Stahl (2004) observed that for dissimilarity matrices with a near-perfect anti-Robinson structure, branching inward tends to perform somewhat better than branching forward. In other words, for well-structured matrices, the bounding process proposed by Defays (1978) is extremely effective and inward branching leads to better pruning of partial solutions. However, the forward algorithm tends to be appreciably more efficient than inward branching for even modest departures from anti-Robinson structure. Because the lipread consonant data are rather messy and devoid of a strong structure, the tremendous advantages for the forward algorithm are not surprising.

11 Seriation—Multiobjective Seriation

11.1 Introduction to Multiobjective Seriation

We now turn our attention to a situation where the quantitative analyst wishes to identify a single permutation that provides a good to excellent fit for each proximity matrix in a family of matrices, C. Alternatively, an analyst might wish to apply various criteria to one or more matrices. Many seriation problems have multiple optima. Therefore, with biobjective programming, the original problem can be formulated to obtain an optimal solution (or near-optimal solution) with the best possible permutation to accommodate a desirable second criterion or matrix. Similarly, a biobjective problem can be extended to a triobjective problem to accommodate a desirable third criterion or matrix.

Brusco and Stahl (2001a) presented a multiobjective dynamic programming approach to seriation that can often accomplish this objective. By developing a scalar-multiobjective-programming-problem (SMPP), their objective was to find an ordering for a single asymmetric proximity matrix that maximized a weighted function of criteria. This is useful for analysts who wish to determine appropriate criterion/criteria such as choosing appropriate gradient indices (equations 9.2.1 through 9.2.4) to maximize. Brusco (2002c) extended the paradigm to seriation of multiple proximity matrices based on a single criterion. He used multiobjective programming to find a permutation that produces good dominance indices for multiple asymmetric matrices, or good anti-Robinson indices for multiple symmetric matrices. The model is readily adapted to incorporate different criteria applied to different matrices (Brusco & Stahl, 2005), such as maximizing the dominance index (equation 8.1) and Defays' criterion (equation 10.2) for two skew-symmetric components of a matrix.

To formulate the multiobjective programming model, as we did in Chapter 6, we assume that optimal permutations and corresponding objective function values, f_q^*, have been obtained for each of the Q matrices in C and/or each criterion considered for the matrix/matrices. We also define user-specified weights, $0 < w_q < 1$ for $q = 1,\ldots, Q$ such that $\sum_{q=1}^{Q} w_q = 1$. If aspiration levels are a concern, then constraints can be added to force individual objective functions to achieve certain values, b_q for each criterion/matrix q. The multiobjective model can then be stated as:

$$\underset{\psi \in \Psi}{Max} : F(\psi) = \sum_{q=1}^{Q} w_q \left(\frac{f_q(\psi)}{f_q^*} \right) = \sum_{q=1}^{Q} \left(\frac{w_q}{f_q^*} \right) f_q(\psi) \qquad (11.1)$$

$$\text{subject to}: \sum_{q=1}^{Q} w_q = 1, \qquad (11.1.1)$$

$$w_q > 0, \qquad (11.1.2)$$

$$f_q(\psi) \geq b_q, \text{ for } q = 1,\ldots,Q. \qquad (11.1.3)$$

Two key concepts must be understood. First, weights for the criteria must be convex (sum to 1) and subsequently normalized by dividing the weight for each criterion by the optimal objective function value for that criterion. Second, when implementing multiobjective seriation, the weights are applied only after calculations for the matrix have been made. There is a temptation to use weighted *rowsums* and a weighted matrix in branch-and-bound implementations to condense programming code. However, this cannot be done without great caution such as with maximizing the dominance index, which can exploit the distributive property of real numbers. However, if such a tactic is used during UDS evaluation, the weights are squared, destroying their convex property. Weights are applied during the individual routines for determining an initial lower bound, the adjacency or interchange test, the bound test, and the evaluation of a permutation for the particular objective function. For a series of Q asymmetric matrices with *normalized* weights (denoted $w(k)$ for $k = 1,\ldots Q$), EVALUATION of a multiobjective seriation problem can be stated algorithmically as:

EVALUATION = 0
for k = 1 to Q
  ObjFnValue = 0
  {If a matrix/criterion has a nonzero weight, then we **determine** the objective function value, *ObjFnValue* for that matrix, $\mathbf{A_k}$, without using

weight. Examples are described in the following sections.}
EVALUATION = EVALUATION + w(k) *ObjFnValue
next k

By placing a small weight on an alternative criterion/matrix, we can often find an optimal (or near-optimal) solution for our original problem with the best possible solution for the alternative criterion/matrix. We also have the options of placing equal weights on all the criteria or using varied weighting schemes to examine the trade-offs among the different criteria.

11.2 Efficient Solutions

In a multiobjective programming problem, we have a series of Q matrices and/or criteria. A permutation of the n objects, ψ, is said to be *dominated* by another permutation, ψ', if and only if $f_q(\psi') \geq f_q(\psi)$ for all objective functions ($q = 1,....,Q$) and $f_q(\psi') > f_q(\psi)$ for at least one objective function. If ψ is not dominated by any alternative permutation ψ', then ψ is said to be *nondominated* or *efficient*. In an important theorem, Soland (1979) proved that any optimal solution to a multiobjective programming problem with strictly positive weights is efficient. Thus, successive trials of SMPP with varying weights will generate an efficient set—also known as the Pareto efficient set, the nondominated set, and efficient frontier. For biobjective (or triobjective) problems, a subset of the efficient frontier can be generated by SMPP and plotted for visual assessment of the trade-off trends between the objective functions.

11.3 Maximizing the Dominance Index for Multiple Matrices

During branch-and-bound procedures for multiobjective seriation, weights are applied after calculating the evaluation of each criterion for each matrix. However, because the dominance index merely sums matrix entries above the diagonal, we can take advantage of the distributive property of the real numbers. As such, prior to performing the branch-and-bound procedure, we can produce a "weighted" matrix and "weighted" *rowsums* to be used throughout the procedure. Let n represent the number of objects and let Q represent the number of matrices. *Rowsums* are assumed to be known for the Q matrices and the weighted *rowsum* for row i is considered the $(Q+1)^{th}$ *rowsum* for the ith row of all matrices. Similarly, the $(Q+1)^{th}$ matrix is the weighted matrix.

```
/* Calculate weighted rowsums */
For i = 1 To n
  rowsum(Q + 1, i) = 0
  For k = 1 To Q
    rowsum(Q + 1, i) = rowsum(Q+1, i) + w(k) * rowsum(k, i)
  Next k
Next i
/* Calculate weighted matrix. */
For i = 1 To n
  For j = 1 To n
    A(Q + 1, i, j) = 0
    For k = 1 To Q
      A(Q + 1, i, j) = A(Q + 1, i, j) + w(k) * A(k, i, j)
    Next k
  Next j
Next i
```

With these weighted *rowsums* and matrix in place, we can simply perform the branch-and-bound algorithm on the $(Q+1)^{th}$ *rowsums* and matrix. The particular routines for the main algorithm—finding the initial lower bound, evaluation, and fathoming—also use these prefabricated totals in the same manner as a single objective problem.

To find an equitable multiobjective solution, we can set each weight for each matrix as $1/Q$, normalize the weights, find the weighted *rowsums* and matrix, and use the branch-and-bound algorithm to solve the multiobjective problem. If we would like to continue finding points in the efficient set, then we can vary the weights, normalize the weights, find the new weighted *rowsums* and matrix, and solve. Not every weighting scheme is unique to a point on the efficient frontier; however, each weighting scheme produces a single point on the efficient frontier. Therefore, a logical methodology is to vary the weight of the matrices over a fixed interval by steady increments. For example, in a biobjective problem, an analyst could vary the first weight by .05 over the interval [.4, .6], i.e. at the starting point plus four steps of .05.

Consider the confusion matrices in Table 11.1 for acoustic recognition of digits according to male or female voice. The maximum dominance index for the male voice *(f1)* is 1751 with an optimal permutation of ψ_M = (5, 9, 1, 4, 7, 6, 8, 3, 2); for the female voice *(f2)*, the maximum dominance index is 4737, indicating more confusion, with a slightly different optimal permutation of ψ_F = (5, 9, 1, 4, 7, 3, 6, 2, 8). The single objective problems correspond to weights of (0, 1) and (1, 0), which do not necessarily generate points on the efficient frontier as per Soland's theorem. (Single objective results are reported and italicized in Table 11.2 merely for convenience and comparison.) In the biobjective problem, we find two points on the efficient frontier when weights on the first matrix are varied from 0 to 1 in 10 increments. Specifically, we find different points for weighting schemes of (0.1, 0.9) and (0.2, 0.8). Closer examination of the weighting schemes about these points yields a third point at (0.08, 0.92), which also corresponds to the optimal solution for *f2*. Figure 11.1 depicts the efficient frontier for values of individual objective functions—*f1* and *f2*—given the optimal biobjective permutations.

Table 11.1. (a and b) Confusion matrices—rows labeled by responses and columns labeled by stimulus—for auditory recognition of the digits 1–9. The data sets are for stimulus delivered by a male voice and stimulus delivered by a female voice as collected by Morgan et al. (1973, p. 376).

	Acoustic Recognition of Digits (Male Voice)								
	(1)	(2)	(3)	(4)	(5)	(6)	(7)	(8)	(9)
(1)	188	23	33	55	21	20	31	21	56
(2)	9	117	60	16	2	15	23	30	6
(3)	12	161	143	20	5	29	31	44	5
(4)	17	33	37	300	7	21	30	36	15
(5)	150	16	20	58	445	11	41	32	113
(6)	8	83	37	10	0	346	38	56	2
(7)	21	51	71	29	5	52	274	59	7
(8)	16	32	60	18	6	27	30	219	11
(9)	119	21	62	29	51	14	29	34	324
	Acoustic Recognition of Digits (Female Voice)								
	(1)	(2)	(3)	(4)	(5)	(6)	(7)	(8)	(9)
(1)	770	96	121	98	128	77	122	114	218
(2)	41	385	224	48	35	94	79	91	60
(3)	56	438	480	54	35	100	85	141	49
(4)	63	89	125	1023	67	63	100	177	61
(5)	254	66	81	119	937	64	83	59	324
(6)	30	110	98	38	16	805	110	242	22
(7)	41	189	168	64	26	191	807	154	51
(8)	64	87	139	58	28	111	109	513	48
(9)	242	96	119	60	292	44	65	55	720

11 Seriation—Multiobjective Seriation

Table11.2. Generating points on the efficient frontier for maximizing the dominance index of matrices in Table 11.1. For a given permutation, the function $f1$ measures the dominance index for the male voice data and $f2$ for the female voice data.

$w(1)$	$w(2)$	$F(\psi^*)$	$f1(\psi^*)$ (% of $f1^*$)	$f2(\psi^*)$ (% of $f2^*$)	Optimal biobjective permutation (ψ^*)
0	1	N/A	1725 (98.5151%)	4737 (100%)	(5, 9, 1, 4, 7, 3, 6, 2, 8)
0.1	0.9	0.9986	1733 (98.972%)	4735 (99.9578%)	(5, 9, 1, 4, 7, 6, 3, 2, 8)
0.2	0.8	0.9986	1751 (100%)	4729 (99.8311%)	(5, 9, 1, 4, 7, 6, 8, 3, 2)
0.3	0.7	0.9986	1751 (100%)	4729 (99.8311%)	(5, 9, 1, 4, 7, 6, 8, 3, 2)
0.4	0.6	0.9988	1751 (100%)	4729 (99.8311%)	(5, 9, 1, 4, 7, 6, 8, 3, 2)
0.5	0.5	0.999	1751 (100%)	4729 (99.8311%)	(5, 9, 1, 4, 7, 6, 8, 3, 2)
0.6	0.4	0.9992	1751 (100%)	4729 (99.8311%)	(5, 9, 1, 4, 7, 6, 8, 3, 2)
0.7	0.3	0.9993	1751 (100%)	4729 (99.8311%)	(5, 9, 1, 4, 7, 6, 8, 3, 2)
0.8	0.2	0.9995	1751 (100%)	4729 (99.8311%)	(5, 9, 1, 4, 7, 6, 8, 3, 2)
0.9	0.1	0.9997	1751 (100%)	4729 (99.8311%)	(5, 9, 1, 4, 7, 6, 8, 3, 2)
1	0	N/A	1751 (100%)	4729 (99.8311%)	(5, 9, 1, 4, 7, 6, 8, 3, 2)
Closer Examination of Efficient Frontier for Weights between (0.08, 0.92) and (0.11, 0.89)					
0.08	0.92	0.9989	1725 (98.5151%)	4737 (100%)	(5, 9, 1, 4, 7, 3, 6, 2, 8)
0.09	0.91	0.9987	1733 (98.972%)	4735 (99.9578%)	(5, 9, 1, 4, 7, 6, 3, 2, 8)
0.1	0.9	0.9986	1733 (98.972%)	4735 (99.9578%)	(5, 9, 1, 4, 7, 6, 3, 2, 8)
0.11	0.89	0.9985	1751 (100%)	4729 (99.8311%)	(5, 9, 1, 4, 7, 6, 8, 3, 2)

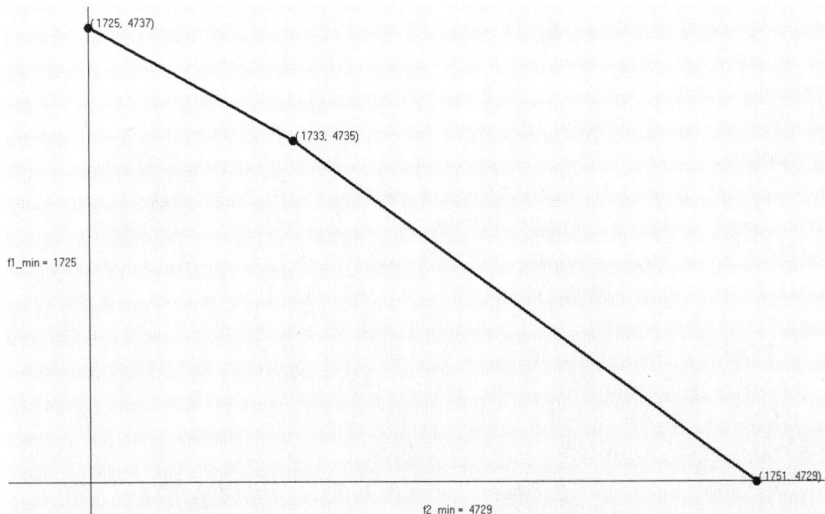

Figure 11.1. Points on the efficient frontier for maximizing the dominance index of matrices in Table 11.1. Axes are defined by single objective solutions.

11.4 UDS for Multiple Symmetric Dissimilarity Matrices

For multiobjective problems in unidimensional scaling, we do not have the luxury of relying on a linear objective function to allow us to merely amalgamate all matrices into one weighted matrix with weighted rowsums. The branch-and-bound procedure remains unaltered; however, the particular routines must be adjusted to perform calculations on all matrices. Once an optimal solution has been found, the coordinates and error are calculated for each matrix according to the optimal permutation. The EVALUATION routine is adjusted as described earlier for multiple matrices. The INTERCHANGETEST must be performed for all matrices. The appropriate pseudocode demonstrates how all matrices are considered in the multiobjective problem.

INTERCHANGETEST = Pass
AltPosition = 1
while (*AltPosition* < *Position* And INTERCHANGETEST)
  CurrentContribution = 0
  AltContribution = 0
  for i = *AltPosition* To *Position*
    for k = 1 to Q
      /* **Determine** SumLeft for *i*th row of matrix $\mathbf{M}_k$ prior to exchanging

positions of objects currently in positions *Position* and *AltPositon*.*/
SumLeft = 0
for j = 1 to i − 1
 SumLeft = SumLeft + **A**(k, *permutation*(i), *permutation*(j))
next j
Contribution = 2 * SumLeft − *rowsum*(k, *permutation*(i))
/* **Accumulate** CurrentContribution for *i*th row of all matrices
subject to their normalize weights for all rows, 1 through *Position*.*/
CurrentContribution = CurrentContribution
                                   + w(k) * Contribution * Contribution
/* **Determine** SumLeft for *i*th row of matrix *k* after exchanging
positions of objects in *Position* and *AltPosition*. */
if (i > *AltPosition* and i < *Position*) then
 SumLeft = SumLeft
                      − **A**(k, *permutation*(i), *permutation*(*AltPosition*))
                      + **A**(k, *permutation*(i), *permutation*(*Position*))
end if
if i = *AltPosition* then
 for j = *AltPosition* + 1 to *Position*
  SumLeft = SumLeft
                      + **A**(k, *permutation*(*AltPosition*), *permutation*(j))
 next j
end if
if i = *Position* then
 for j = *AltPosition* to *Position* − 1
  SumLeft = SumLeft
                      − **A**(k, *permutation*(*Position*), *permutation*(j))
 next j
end If
Contribution = 2 * SumLeft − *rowsum*(k, *permutation*(i))
/* **Accumulate** alternate contribution for *i*th row of all matrices
subject to their normalized weights for all rows, 1,..., *Position*. */
AltContribution = AltContribution
                                   + w(k) * Contribution * Contribution
 next k /* **Perform** calculations for next matrix. */
next i
AltPosition = AltPosition + 1
if CurrentContribution < AltContribution then
 INTERCHANGETEST = Fail
end if
loop /* AltPosition loop */

11.4 UDS for Multiple Symmetric Dissimilarity Matrices

Similarly, we adjust the BOUNDTEST to account for all matrices. The matrix **B**, containing maximum possible contributions to the objective function for each object in each position, is once again calculated and stored immediately after the initial lower bound is found. We assume that the *rowsums* have been previously determined.

```
/* Initialize B */
for i = 1 to n
 for j = 1 to n
 B(i, j) = 0
 next j
next i
/* Determine matrix for UpperB2 */
for object = 1 to n
 /* Order (low-to-high) entries in row (object) for each matrix */
 for k = 1 to Q
  for Index = 1 to n    /* deleting diagonal entry from Order */
   if Index < object then Order(k, Index) = A(k, object, Index)
   if Index > object then Order(k, Index – 1) = A(k, object, Index)
  next Index
  for i = 1 to n – 1
   Minimum = Order(k, i)
   for j = i + 1 to n – 1
    if Order(k, j) < Minimum then
     Minimum = Order(k, j), Order(k, j) = Order(k, i),
     Order(k, i) = Minimum
    end if
   next j
  next i
 next k
 /* Calculate $b_{ij}$ for all positions of object in permutation */
 for Position = 1 to n
  SumLeft = 0
  if Position < ((n / 2) + 1) then
   B(object, Position) = 0
   for k = 1 to Q
   SumLeft = 0
   for i = 1 to Position – 1
    SumLeft = SumLeft + Order(k, i)
   next i
   SumRight = rowsum(k, object) – SumLeft
```

      B(object, *Position*) = **B**(object, *Position*)
                + $w(k)$ * (SumRight − SumLeft) * (SumRight − SumLeft)
   next k
  else
   B(object, *Position*) = **B**(object, n − *Position* + 1)
  end if
 next *Position*
next object

Notice that the *rowsums*, which are determined prior to finding the initial lower bound, are found for *each* row of *each* matrix. The entries of the matrix **B** accumulate maximum contributions to the overall objective function by *each* matrix with a particular object in a particular position according to the normalized convex weights. Now, for the BOUNDTEST, we can quickly determine an upper bound for a partial sequence.

UpperB1 = 0: *UpperB2* = 0
/* **Determine** S\R(*Position*). */
Index = *Position* + 1
for i = 1 to n
 found = False
 for j = 1 to *Position*
  if *permutation*(j) = i then found = True
 next j
 if not found then
  complement(Index) = i
  Index = Index + 1
 end if
next i
for k = 1 to Q
 for i = 1 to *Position*
  temp = 0
  for j = 1 to i − 1
   temp = temp + **A**(k, *permutation*(i), *permutation*(j))
  next j
  temp = *rowsum*(k, *permutation*(i)) − 2 * temp
  temp = $w(k)$ * temp * temp
  UpperB1 = UpperB1 + temp
 next i
/* **Find** maximum contribution to objective function value by the complement. */

11.4 UDS for Multiple Symmetric Dissimilarity Matrices

```
    for UnfilledPosition = Position + 1 to n
    Maximum = B(complement(Position + 1), UnfilledPosition)
    for UnselectedObject = Position + 2 to n
      if B(complement(UnselectedObject), UnfilledPosition) > Maximum
      then Maximum = B(complement(UnselectedObject), UnfilledPosition)
    next UnselectedObject
    UpperB2 = UpperB2 + Maximum
    next UnfilledPosition
next k
UpperB = UpperB1 + UpperB2
```

Recalling the examples of unidimensional scaling from Chapter 10, we can take a closer look at the seemingly disagreeable optimal permutations and coordinates for the two closely related data sets. In Table 11.3, we can see the migration from one optimal permutation to the other. Optimal solutions for single objective problems are italicized in Table 11.3 and the single objective solutions assist in defining the axes in Figure 11.2. Specifically, the objective function value for the secondary criterion is at its minimum when the primary criterion is optimized in a single objective problem. This is the reason that efficient frontier does not necessarily "touch" the axis, as in the case of the point (1201, 1251) for weighting scheme (1, 0). Once again, single objective solutions are not necessarily members of the efficient set. However, the point for the weighting scheme (0.05, 0.95) is the same as the optimal single objective solution for the second criteria ($g2$), which extends the efficient frontier to the minimum axis for criterion $g1$.

This efficient frontier has several efficient points. Notice that a minor change in weights can appear to "flip" the permutation, which is actually an artifact of the symmetry flags *SymFirst* and *SymSecond*; see the migration from the weighting scheme (0.45, 0.55) to (0.5, 0.5) and again from (0.65, 0.35) to (0.7, 0.3). Also notice the trend in the value of $G(\psi^*)$ which decreases until the weighting scheme of (0.65, 0.35) and then increases until the near-optimal solution for $g1$. The decreasing trend does not reverse at equal convex weights of (0.5, 0.5), indicating that the matrix in Table 10.3 with greater confusion has a strong pull on the less confusing matrix of Table 10.4.

11 Seriation—Multiobjective Seriation

Table 11.3. Points generated for the efficient frontier when applying biobjective UDS to the confusion data for consonants degraded by –18db and –12db (Miller & Nicely, 1955). For a given permutation, $g1$ measures the Defays criterion for the –18db data and $g2$ for the –12db data.

$w(1)$	$w(2)$	$G(\psi^*)$	$g1(\psi^*)$ (% of $f1^*$)	$g2(\psi^*)$ (% of $f2^*$)	Optimal biobjective permutation (ψ^*)
0	1	N/A	1192.63982252 (99.2318%)	1270.4331286 (100%)	(t, k, p, f, θ, s, ʃ, 3, z, d, g, ð, v, b, m, n)
0.05	0.95	0.9999	1192.63982252 (99.2318%)	1270.4331286 (100%)	(t, k, p, f, θ, s, ʃ, 3, z, d, g, ð, v, b, m, n)
0.1	0.9	0.9996	1192.8183188 (99.2467%)	1270.41475884 (99.9986%)	(t, k, p, f, θ, ʃ, s, 3, z, d, g, ð, v, b, m, n)
0.15	0.85	0.9992	1193.25412256 (99.283%)	1270.3402682 (99.9927%)	(t, p, k, f, θ, ʃ, s, 3, z, d, g, ð, v, b, m, n)
0.2	0.8	0.9989	1193.25412256 (99.283%)	1270.3402682 (99.9927%)	(t, p, k, f, θ, ʃ, s, 3, z, d, g, ð, v, b, m, n)
0.25	0.75	0.9985	1194.04238284 (99.3485%)	1270.07806468 (99.9721%)	(t, p, k, f, θ, ʃ, s, 3, z, d, g, v, b, ð, m, n)
0.3	0.7	0.9982	1194.04238284 (99.3485%)	1270.07806468 (99.9721%)	(t, p, k, f, θ, ʃ, s, 3, z, d, g, v, b, ð, m, n)
0.35	0.65	0.9978	1194.04238284 (99.3485%)	1270.07806468 (99.9721%)	(t, p, k, f, θ, ʃ, s, 3, z, d, g, v, b, ð, m, n)
0.4	0.6	0.9975	1194.39919652 (99.3782%)	1269.86984556 (99.9557%)	(p, t, k, f, θ, ʃ, s, 3, z, d, g, v, b, ð, m, n)
0.45	0.55	0.9972	1194.39919652 (99.3782%)	1269.86984556 (99.9557%)	(p, t, k, f, θ, ʃ, s, 3, z, d, g, v, b, ð, m, n)
0.5	0.5	0.997	1196.808673 (99.5787%)	1267.71883456 (99.7863%)	(3, z, g, d, ð, b, v, n, m, s, ʃ, θ, f, k, t, p)
0.55	0.45	0.9968	1196.808673 (99.5787%)	1267.71883456 (99.7863%)	(3, z, g, d, ð, b, v, n, m, s, ʃ, θ, f, k, t, p)
0.6	0.4	0.9967	1198.42538244 (99.7132%)	1265.46336088 (99.6088%)	(3, z, g, d, ð, b, v, n, m, θ, f, p, k, t, s, ʃ)
0.65	0.35	0.9967	1198.42538244 (99.7132%)	1265.46336088 (99.6088%)	(3, z, g, d, ð, b, v, n, m, θ, f, p, k, t, s, ʃ)
0.7	0.3	0.9968	1199.2132036 (99.7788%)	1263.85166888 (99.482%)	(θ, f, p, t, k, ʃ, s, m, n, v, b, ð, d, z, g, 3)
0.75	0.25	0.9969	1200.647193 (99.8981%)	1259.6138228 (99.1484%)	(ʃ, f, p, t, k, s, θ, m, n, v, b, d, ð, z, g, 3)
0.8	0.2	0.9971	1201.51024336 (99.9699%)	1256.62626288 (98.9132%)	(ʃ, f, p, t, k, s, θ, m, n, d, v, b, z, ð, g, 3)

11.4 UDS for Multiple Symmetric Dissimilarity Matrices

Table 11.3. Continued

w(1)	w(2)	G(ψ*)	g1(ψ*)	g2(ψ*)	Optimal biobjective permutation (ψ*)
0.85	0.15	0.9976	1201.51024336 (99.9699%)	1256.62626288 (98.9132%)	(3, g, δ, z, b, v, d, n, m, θ, s, k, t, p, f, ʃ)
0.9	0.1	0.9981	1201.73896008 (99.9889%)	1254.83793108 (98.7725%)	(ʃ, f, p, t, k, s, θ, m, n, d, v, z, b, g, δ, 3)
0.95	0.05	0.9987	1201.84719276 (99.9979%)	1253.26369116 (98.6485%)	(ʃ, f, p, θ, t, k, s, m, n, d, v, z, b, g, δ, 3)
1	0	N/A	*1201.87214164 (100%)*	*1251.49216996 (98.5091%)*	*(ʃ f, p, θ, s, k, t, m, n, d, z, v, b, g, δ, 3)*

To plot the points of the efficient set, we evaluate a criterion with the optimal permutation for the other criterion. Also, we note that some points are extremely close for one or both of the criteria, such as (1201.739, 1254.838) and (1201.8472, 1253.2637). Unfortunately, this can necessitate referring to Table 11.3 for exact values. Nevertheless, the general plot illustrates the migration from one optimal solution to the other with more fluidity than a table of extremely close values.

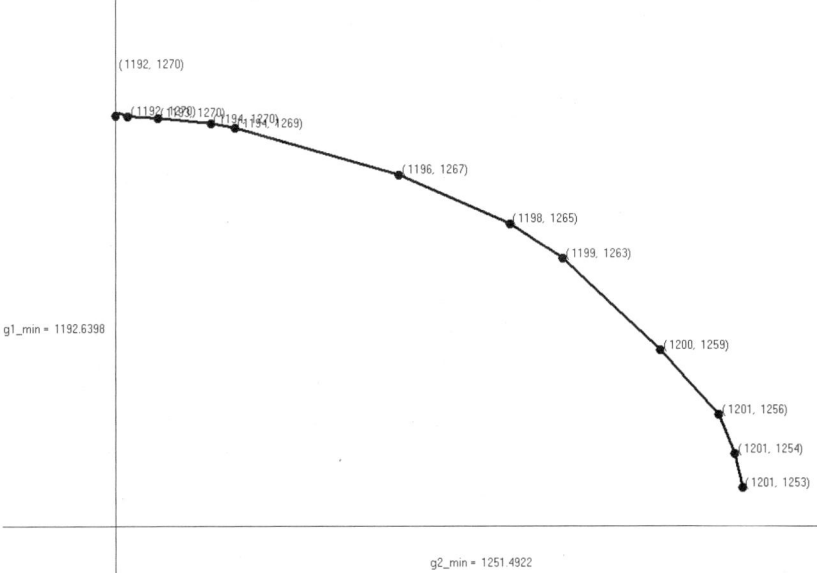

Figure 11.2. The efficient frontier for multiobjective unidimensional scaling of the acoustically degraded consonant data of Tables 10.3 and 10.4. Axes are defined by the values of the alternate criterion when evaluated with the optimal permutation for a primary criterion.

11.5 Comparing Gradient Indices for a Symmetric Dissimilarity Matrix

When we wish to analyze a single matrix with multiple criteria, the main algorithm and the calculation for the initial lower bound remain the same. However, the routines called by these algorithms must be performed with respect to the desired weighting scheme. The main algorithm calls to evaluation and fathoming routines; the calculation of the initial lower bound uses the evaluation routine.

We adopt the convention of using EVALUATION1 as the evaluation routine for the first gradient index ($f1 = U_r$), EVALUATION2 as the evaluation routine for the second gradient index criteria, and so forth. The overall EVALUATION routine uses normalized weights derived from a convex combination of weights. Therefore, the overall objective function value will actually be a percentage, i.e. a value between 0 and 1 inclusive. Conditional statements augment implementation by avoiding computation of irrelevant criteria for a particular problem.

EVALUATION = 0
if $w(1) > 0$ then
  EVALUATION = $w(1)$ * EVALUATION1
if $w(2) > 0$ then
  EVALUATION = EVALUATION + $w(2)$ * EVALUATION2
if $w(3) > 0$ then
  EVALUATION = EVALUATION + $w(3)$ * EVALUATION3
if $w(4) > 0$ then
  EVALUATION = EVALUATION + $w(4)$ * EVALUATION4

In the case of maximizing gradient indices, the fathoming routines are the adjacency test and the bound test. For all of the gradient indices, the adjacency test is based on an inequality. The example of algorithmic pseudocode in section 9.2.2 used *compare1* and *compare2* variables to evaluate the left- and right-hand sides of the equation. If we are interested in multiple gradient indices, then we accumulate the left- and right-hand sides of the equation with respect to the normalized weights (assuming that *compare1* and *compare2* are global variables).

if *Position* = 1 then
  ADJACENCYTEST = Pass
else
  Set Lefthand = 0 and Righthand = 0.
  if $w(1) > 0$ then

```
    ADJACENCYTEST1
    Lefthand = w(1) * compare1
    Righthand = w(1) * compare2
  end if
  if w(2) > 0 then
    ADJACENCYTEST2
    Lefthand = Lefthand + w(2) * compare1
    Righthand = Righthand + w(2) * compare2
  end If
  if w(3) > 0 then
    ADJACENCYTEST3
    Lefthand = Lefthand + w(3) * compare1
    Righthand = Righthand + w(3) * compare2
  end if
  if w(4) > 0 then
    ADJACENCYTEST4
    Lefthand = Lefthand + w(4) * compare1
    Righthand = Righthand + w(4) * compare2
  end if
  if Lefthand < Righthand then
    ADJACENCYTEST = False
  else
    ADJACENCYTEST = True
  end if
end if
```

Similarly, we rely on the fact that all bound tests calculate an upper bound to which the current lower bound is compared, as explained in section 9.2.3. By calculating the overall upper bound with respect to normalized weights (assuming *UpperB* is a global variable), we can determine the feasibility of a partial sequence as a candidate for an optimal permutation for the overall objective function.

```
If Position < 3 then
    BOUNDTEST = Pass
Else
    BT = 0
    if w(1) > 0 then
      BOUNDTEST1
      BT = w(1) * UpperB1
    end if
    if w(2) > 0 then
```

```
    BOUNDTEST2
    BT = w(2) * UpperB2
  end if
  if w(3) > 0 then
    BOUNDTEST3
    BT = w(3) * UpperB3
  end if
  if w(4) > 0 then
    BOUNDTEST4
    BT = w(1) * UpperB4
  end if
  if BT < LowerB then
    BOUNDTEST = Fail
  else
    BOUNDTEST = Pass
  end if
end if
```

Recall the archaeological example of Chapter 9. We have varying optimal permutations for the four gradient indices. An interesting problem is to compare results for the weighted within row gradient and the unweighted within row gradient for the archaeological data in Table 9.1. Referring to the gradient indices, U_r and W_r, we can vary the weights for the individual criteria from (0.05, 0.95) to (0.95, 0.05) for finding points on the efficient frontier for F. Points found on the efficient frontier in this migration are reported in Table 11.4. For use as reference to endpoints and to illustrate the usefulness of the biobjective modeling, the single objective solutions are italicized above and below the points of the efficient frontier. We stress that a requirement of Soland's theorem is strictly positive weights for efficient solutions. This subset of the efficient frontier has six points—(549, 23844), (551, 23837), (553, 23826), (557, 23773), (561, 23651), and (562, 23437). These six points are plotted in Figure 11.3.1. The first important observation is in the weighting scheme of (0.85, 0.95), which finds an alternative optimal solution for the first criteria that improves the value of the second criterion as promised in Chapter 9. This clearly demonstrates the usefulness of biobjective programming when a second criterion is considered but not of great importance. In addition, we can see how the trade-offs become more drastic as we approach the single objective solution for Criteria 1 (U_r), as when moving from the fifth point to the sixth point there is a mere increase of 1 in $f1^*$ yet a drop of 214 in $f3^*$.

Table 11.4. Biobjective seriation of the Kabah collection by weighted and unweighted within row gradient indices to find points in the efficient set.

w(1)	w(2)	F(ψ*)	$f1(\psi^*)$ (% of $f1^*$)	$f3(\psi^*)$ (% of $f3^*$)	Optimal biobjective permutation (ψ*)
0	1	N/A	549 (97.6868%)	23844 (100%)	(IVB, III, VIA, VIB, VA, VB, IVA, VIII, IX, XA, XB, IB, XIB, XIA, IA, VII, II)
0.05	0.95	0.9988	549 (97.6868%)	23844 (100%)	(IVB, III, VIA, VIB, VA, VB, IVA, VIII, IX, XA, XB, IB, XIB, XIA, IA, VII, II)
0.1	0.9	0.9978	551 (98.0427%)	23837 (99.9706%)	(IVB, III, VIA, VIB, VA, VB, IVA, VIII, XA, IX, XB, IB, XIB, XIA, IA, VII, II)
0.15	0.85	0.997	553 (98.3986%)	23826 (99.9245%)	(III, IVB, VIA, VIB, VA, VB, IVA, VIII, XA, IX, XB, IB, XIB, XIA, IA, VII, II)
0.25	0.75	0.9955	557 (99.1103%)	23773 (99.7022%)	(III, IVB, VIB, VIA, VA, VB, IVA, VIII, XA, IX, XB, IB, XIB, XIA, IA, VII, II)
0.45	0.55	0.9947	561 (99.8221%)	23651 (99.1906%)	(VIA, VIB, VA, III, IVB, VB, IVA, VIII, XA, IX, XB, IB, XIB, XIA, IA, VII, II)
0.85	0.15	0.9974	562 (100%)	23437 (98.2931%)	(III, IVB, VA, VIB, VB, VIA, IVA, VIII, XA, IX, XB, IB, XIB, XIA, IA, VII, II)
1	0	N/A	562 (100%)	23429 (98.2595%)	(III, IVB, VA, VIB, VB, VIA, IVA, VIII, XA, IX, IB, XB, XIB, XIA, IA, VII, II)

We can also take a closer look at the nature of the within row gradients and the within row and column gradients as shown in Table 11.5. We can see that any weight on the first criteria (U_r) forces the second criteria (U_{rc}) to find an alternative optimal solution in near-reverse order. On the opposite end of the efficient frontier, the smallest weight on the second criteria forces the first criteria to find an alternative optimal solution by simply reversing the placements of objects XB and IB. In both cases, the primary criterion remains optimal while improving the value for the secondary criterion. For the selected weighting schemes, the biobjective program generates four points on the efficient frontier as shown in Figure 11.3.2. This figure graphically demonstrates the improvement in $f1$ (a 10% improvement) simply by finding an alternate optimal permutation for the second criterion using the biobjective program.

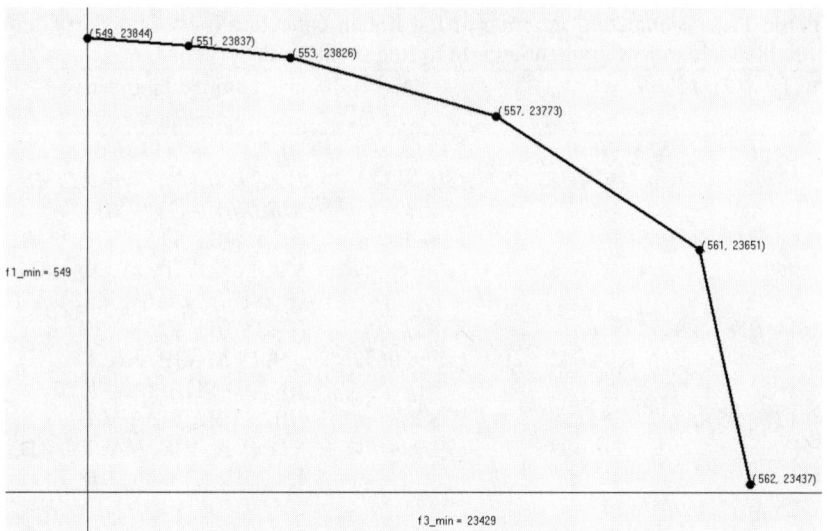

Figure 11.3.1. Plotting a subset of the efficient frontier for maximizing Criteria 1 (U_r) and Criteria 3 (W_r) applied to the data for the Kabah collection.

11.6 Multiple Matrices with Multiple Criteria

We begin by noting that a matrix can be decomposed into its symmetric and skew-symmetric components (Tobler, 1976). Moreover, the skew-symmetric component can be factored according to its magnitude (absolute value) and direction (sign):

$$A_q = \{(a_{ij} + a_{ji})/2\} + \{(a_{ij} - a_{ji})/2\} \tag{11.2}$$
$$= \{(a_{ij} + a_{ji})/2\} + \{|(a_{ij} - a_{ji})/2|\} * \{sign((a_{ij} - a_{ji})/2)\}^T$$

Although the temptation might be to apply Defays' maximization to the symmetric component and optimize the dominance index for the skew-symmetric component, many proximity matrices have a rather uninteresting and uninformative symmetric component, such as matrices constructed to reflect proportions or percentages that inherently yield $a_{ij} = 1 - a_{ji}$ (so that $(a_{ij} + a_{ji})/2 = (1 - a_{ji} + a_{ji})/2 = ½$ for all entries of the symmetric matrix). However, conjoint seriation of the magnitude and direction of the skew-symmetric component can be illuminating (Brusco & Stahl, 2005). In this multiobjective programming problem, we use two different matrices each being ordered according to different criteria. Spe-

11.6 Multiple Matrices with Multiple Criteria

cifically, we order the magnitude of entries as per Defays' maximization (equation (8.1)) and we order the sign elements to optimize the dominance index (equation (10.2)). Our objective function thus becomes:

$$\max_{\psi \in \Psi} : F(\psi) = w_f \left(\frac{f(\psi)}{f^*} \right) + w_g \left(\frac{g(\psi)}{g^*} \right) = \left(\frac{w_f}{f^*} \right) f(\psi) + \left(\frac{w_g}{g^*} \right) g(\psi). \quad (11.3)$$

To implement a branch-and-bound process to find the optimal solution for (11.3), we follow the now familiar pattern of formulating *joint* routines for evaluation and fathoming using normalized weights on the relevant outcomes of individual routines. Therefore, for this multiobjective problem, we find an initial lower bound by using only the highest-to-lowest ranking of weighted graduated *rowsums* and apply the approach to a local optimum. (Note: We are using *weighted* graduated *rowsums* merely to find our initial *LowerB*.) Also, after the initial lower bound is found and prior to starting the main algorithm, we must calculate and store the **B** matrix for the second component of the bound test to unidimensionally scale the magnitude matrix. As described earlier, the EVALUATION and BOUNDTEST use sums of weighted individual evaluations and upper bounds. Two procedures, the symmetry check and the adjacency/interchange test, deviate slightly from earlier descriptions. Because maximizing the dominance index does not take advantage of symmetry the way that unidimensional scaling does, the "direction" of the optimal permutation is determined by the sign matrix. Hence, no symmetry check is needed in the main algorithm. Until now, we have used an adjacency test for maximizing the dominance index and an interchange test for unidimensional scaling. Therefore, we will "upgrade" the adjacency test for maximizing the dominance index to an interchange test.

To understand an interchange test for maximizing the dominance index, recall Figure 10.1. The objective function for the dominance index only sums the entries in the upper triangle, i.e., the entries to the right of the diagonal in each row. Our interchange test thus becomes a matter of satisfying equation (11.4):

$$\sum_{i=q+1}^{p} a_{\psi(q)\psi(i)} + \sum_{i=q+1}^{p-1} a_{\psi(i)\psi(p)} \geq \sum_{i=q}^{p-1} a_{\psi(p)\psi(i)} + \sum_{i=q+1}^{p-1} a_{\psi(i)\psi(q)}. \quad (11.4)$$

166   11 Seriation—Multiobjective Seriation

Table 11.5. A subset of the efficient set for maximizing unweighted within row gradient indices and unweighted within row and column gradient indices for the Kabah collection.

$w(1)$	$w(2)$	$F(\psi^*)$	$f1(\psi^*)$ (% of $f1^*$)	$f2(\psi^*)$ (% of $f2^*$)	Optimal biobjective permutation (ψ^*)
0	1	N/A	499 (88.79%)	1050 (100%)	(II, VII, IA, XIA, XIB, IB, XB, IX, XA, VIII, IVA, VB, VA, VIB, VIA, III, IVB)
0.05	0.95	0.9992	553 (98.3986%)	1050 (100%)	(III, IVB, VIA, VIB, VA, VB, IVA, VIII, XA, IX, XB, IB, XIB, XIA, IA, VII, II)
0.7	0.3	0.9898	557 (99.1103%)	1036 (98.6667%)	(III, IVB, VIB, VIA, VA, VB, IVA, VIII, XA, IX, XB, IB, XIB, XIA, IA, VII, II)
0.85	0.15	0.9905	561 (99.8221%)	994 (94.6667%)	(VIA, VIB, VA, III, IVB, VB, IVA, VIII, XA, IX, XB, IB, XIB, XIA, IA, VII, II)
0.9	0.1	0.9932	562 (100%)	979 (93.2381%)	(III, IVB, VA, VIB, VB, VIA, IVA, VIII, XA, IX, XB, IB, XIB, XIA, IA, VII, II)
1	0	N/A	562 (100%)	976 (92.9524%)	(III, IVB, VA, VIB, VB, VIA, IVA, VIII, XA, IX, IB, XB, XIB, XIA, IA, VII, II)

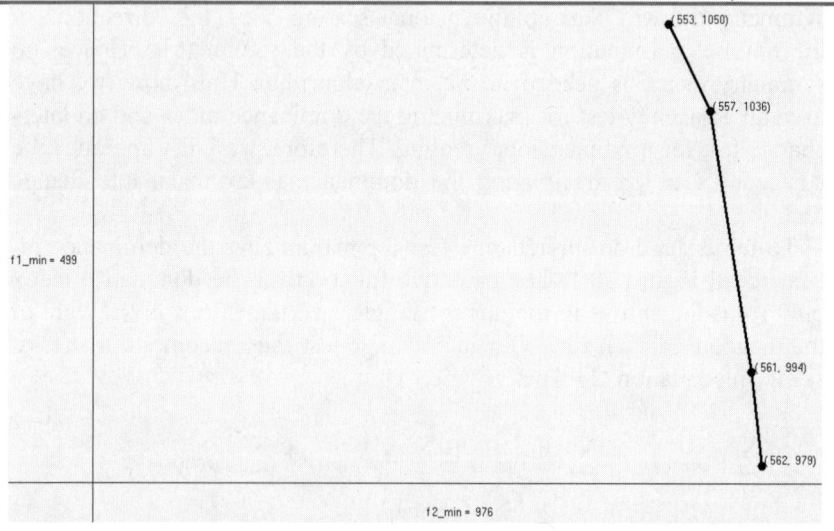

Figure 11.3.2. Four points on the efficient frontier for maximizing Criteria 1 (U_r) and Criteria 2 (U_{rc}) as applied to the Kabah collection.

11.6 Multiple Matrices with Multiple Criteria

The INTERCHANGETEST for unidimensional scaling remains the same as for a single objective UDS problem. Using matrices **Sign** and **Magnitude** to hold the direction and magnitude information of the skew-symmetric component of the proximity matrix with respective normalized weights of $w(s)$ and $w(m)$, we can write our algorithmic pseudocode for the biobjective interchange test.

```
if Position = 1 then
  INTERCHANGETEST = Pass
else
  INTERCHANGETEST = Pass
  AltPosition = 1
  while (AltPosition < Position and INTERCHANGETEST)
    CurrentContribution = 0
    AltContribution = 0
    if w(m) > 0 then
    /* Calculate contribution of Unidimensional Scaling function
       for rows permutation(AltPosition) through permutation(Position),
       one at a time */
    for i = AltPosition to Position
      /* Calculate current contribution, before swapping positions of
         objects in permutation(Position) and permutation(AltPosition) */
      SumLeft = 0
      for j = 1 to i − 1
        SumLeft = SumLeft + Magnitude(permutation(i), permutation(j))
      next j
      UDSContribution = 2 * SumLeft - rowsum(1, permutation(i))
      CurrentContribution = CurrentContribution
           + w(magnitude) * UDSContribution * UDSContribution
      /* Calculate alternate contribution, after swapping positions of
         objects in permutation(Position) and permutation(AltPosition) */
      if (i > AltPosition and i < Position) then
        SumLeft = SumLeft
           − Magnitude(permutation(i), permutation(AltPosition))
           + Magnitude(permutation(i), permutation(Position))
      end if
      if i = AltPosition then
        for j = AltPosition + 1 to Position
          SumLeft = SumLeft
             + Magnitude(permutation(AltPosition), permutation(j))
        next j
```

```
      end if
      if i = Position then
         for j = AltPosition to Position − 1
           SumLeft = SumLeft
                − Magnitude (permutation(Position), permutation(j))
         next j
      end if
      UDSContribution = 2 * SumLeft − rowsum(m, permutation(i))
      AltContribution = AltContribution
                + w(m) * UDSContribution * UDSContribution
    next i
   end if
   if w(2) > 0 then
      /* Calculate contribution of Dominance Index function for rows
      permutation(AltPosition) through permutation(Position) */
      for i = AltPosition + 1 To Position
         CurrentContribution = CurrentContribution
              + w(s) * Sign(permutation(AltPosition), permutation(i))
      next i
      for i = AltPosition + 1 to Position − 1
         CurrentContribution = CurrentContribution
                + w(s) * Sign(permutation(i), permutation(Position))
         AltContribution = AltContribution
                + w(s) * Sign(permutation(i), permutation(AltPosition))
      next i
      for i = AltPosition to Position − 1
         AltContribution = AltContribution
                + w(s) * Sign(permutation(Position), permutation(i))
      next i
   end if
   AltPosition = AltPosition + 1
   if CurrentContribution < AltContribution then
      INTERCHANGETEST = Fail
  loop   /* AltPosition loop */
end if
```

As an example, we return to the matrix for assessment of criminal offences (Table 8.6). The results in Table 11.6 agree with the results obtained by Brusco and Stahl (2005) for the same biobjective problem when employing dynamic programming to find optimal permutations.

Table 11.6. A subset of the efficient set for maximizing the dominance index for the magnitude component and the Defays' maximization for the magnitude component of the skew-symmetric component of Table 8.6.

$w(1)$	$w(2)$	$F(\psi^*)$	$f(\psi^*)$ (% of f^*)	$g(\psi^*)$ (% of g^*)	Optimal biobjective permutation (ψ^*)
0	1	N/A	95 (94.0594%)	122813150.5 (100%)	(Libel, Larceny, Burglary, Assault & Battery, Forgery, Counterfeiting, Perjury, Embezzlement, Arson, Adultery, Kidnapping, Seduction, Abortion, Homicide, Rape)
0.01	0.99	.9996	97 (96.0396%)	122806940.5 (99.9949%)	(Libel, Larceny, Assault & Battery, Burglary, Forgery, Counterfeiting, Perjury, Embezzlement, Arson, Adultery, Kidnapping, Seduction, Abortion, Homicide, Rape)
0.05	0.95	.9984	99 (98.0198%)	122733072.5 (99.9348%)	(Libel, Larceny, Assault & Battery, Burglary, Forgery, Counterfeiting, Perjury, Embezzlement, Arson, Adultery, Kidnapping, Seduction, Abortion, Rape, Homicide)
1	0	N/A	101 (100%)	122562702.5 (99.7961%)	(Libel, Larceny, Assault & Battery, Burglary, Forgery, Counterfeiting, Perjury, Embezzlement, Arson, Kidnapping, Adultery, Seduction, Abortion, Rape, Homicide)

11.7 Strengths and Limitations

The strengths and limitations of a multiobjective programming approach to seriation are similar to those for multiobjective clustering. Among the advantages of multiobjective programming are favorable tie-breaking in the case of alternative optima, additional confidence when a single permutation produces excellent values for multiple criteria, and, in the case of multiple data sources, preventing one source from having too profound an impact on the solution obtained. The principal disadvantage, of course, is the selection of appropriate weights, and this problem is exacerbated as the number of relevant criteria increases. The branch-and-bound approach is easily adapted to accommodate multiple objectives, and examination of the efficient frontier can assist an analyst in determining appropriate weights.

Part III

Variable Selection

12 Introduction to Branch-and-Bound Methods for Variable Selection

12.1 Background

In many applications, analysts have a large set of candidate variables ($V = \{v_1, v_2, ..., v_D\}$) for possible inclusion in a multivariate analysis of statistical data, but wish to select only a subset of d variables, $V_d \subset V$. Areas of multivariate analysis for which this type of selection problem is often relevant include regression analysis, discriminant analysis, and cluster analysis. The problem of selecting a subset of variables for these and other applications has been recognized in several literature bases. In the psychometric and statistical literature, the problem is generally referred to as *variable selection*, whereas the machine learning literature favors the term *feature selection*. Throughout Chapters 12 through 14, we will use the term variable selection, but this does not in any way restrict the applicability of the methods to particular fields or disciplines.

Although we could approach the variable-selection problem via exhaustive enumeration of all possible subsets of variables (Furnival, 1971), this is often a computationally expensive undertaking and frequently unnecessary. The acceleration of the search process can be accomplished through the use of heuristic procedures; however, the selection of an optimal subset is no longer guaranteed. Branch-and-bound, on the other hand, accelerates the search process while preserving the identification of an optimal subset. Branch-and-bound approaches for variable selection have been implemented since at least the late 1960s (Beale, Kendall, & Mann, 1967; Furnival & Wilson, 1974; Hamamoto, Uchimura, Matsura, Kanaoka, & Tomita, 1990; Hocking & Leslie, 1967; Kittler, 1978; LaMotte & Hocking, 1970; Miller, 2002, Chapter 3; Narendra & Fukunaga, 1977), and the potential areas of application are numerous. Narendra and Fukunaga (1977) developed a branch-and-bound algorithm for selection of a subset of d variables from a set of size D, with a general objective of maximizing a quadratic form associated with a real symmetric matrix. Of particular interest were functions of the following structure:

$$\Omega(V_d) = \mathbf{x}'\mathbf{\Lambda}^{-1}\mathbf{x}, \tag{12.1}$$

where $V_d \subset V$ is a subset of d selected variables, $\mathbf{\Lambda}$ is a $d \times d$ matrix corresponding to the selected variables, $\mathbf{x}$ is a $d \times 1$ vector of measurements, and $\mathbf{x}'$ is the transpose of $\mathbf{x}$.

One example where (12.1) has particular relevance is when $\mathbf{x}$ is a vector of variable differences between a pair of objects and $\mathbf{\Lambda}$ is a covariance matrix. Under these conditions, (12.1) corresponds to Mahalanobis' distance. Another criterion with this class of functions is the Fisher criterion.

Roberts (1984) published a Fortran program for (12.1), which was subsequently corrected by Cobby (1986, 1991). Ridout (1988) also published a Fortran program for (12.1), which incorporates an important sorting step that increases computational efficiency. Discussions of applications of (12.1) can be collected from various pattern recognition textbooks (Bishop, 1995; Fukunaga, 1990).

The crux of criterion (12.1) is that it adheres to what Narendra and Fukunaga (1977), Hand (1981a), Roberts (1984), Ridout (1988), Bishop (1995), and others refer to as a *monotonicity* property. Assuming that the optimization problem is of the maximization variety, the monotonicity property implies that if adding a variable modifies a given subset, then the resulting clustering criterion must improve or stay the same. More formally, the property implies that for any $V_d \subset V_{d+1}$, the following condition must hold:

$$\Omega(V_d) \leq \Omega(V_{d+1}). \tag{12.2}$$

The monotonicity property is also relevant to applications of the branch-and-bound paradigm to variable selection problems in regression (Furnival & Wilson, 1974). In these applications, the objective criterion is typically to minimize the residual sum of squares error for a fixed subset size. However, the behavior of this criterion is such that the modification of the subset by removing (as opposed to adding) a variable cannot possibly improve the objective function. Thus, the monotonicity property is still applicable, but from a slightly different perspective.

One of the disadvantages of the monotonicity property is that it entails difficulty when comparing subsets of different sizes. Referring again to the original context of (12.1) under a maximization objective, an optimally selected subset of $d + 1$ variables should produce a better solution than an optimally selected subset of d variables. However, the question as to whether the improvement realized from increasing the subset size is worthwhile is rather subjective. Despite the fact that the monotonicity property creates difficulty with respect to comparison of subsets of different sizes, it facilitates the implementation of a branch-and-bound approach for finding an optimal subset for a fixed size.

The following two chapters present branch-and-bound applications for variable selection. Chapter 13 focuses on variable selection in cluster analysis, whereas Chapter 14 considers variable selection in regression analysis. Both of these application areas are of crucial importance. Although they are somewhat similar in structure, the algorithm for cluster analysis begins with an empty set of selected variables and adds variables to the set during the implementation of the algorithm. On the other hand, the branch-and-bound regression algorithm begins with the complete set of variables as the selected set and discards variables during the implementation. For this reason, we have opted to present (independently) the branch-and-bound algorithms for these two areas in their respective chapters, rather than offer a general branch-and-bound paradigm here.

13 Variable Selection for Cluster Analysis

13.1 *True* Variables and *Masking* Variables

Within the context of cluster analysis applications, analysts often focus on the fact that among the set of D candidate clustering variables ($V = \{v_1, v_2, ..., v_D\}$) for possible for inclusion in a cluster analysis, only a subset, $V_T \subseteq V$, of those variables might be appropriate for uncovering structure in the data set. We refer to variables in V^T as *true variables* because they define a true structure in the data set. Unfortunately, incorporating the entire set of candidate variables, V, into the multivariate analysis is generally ineffective because the inclusion of the irrelevant variables (those in V but not V^T, $V \setminus V_T$) impedes the recovery of the true cluster structure. In other words, the irrelevant variables hide or obfuscate the true structure in the data set. This led Fowlkes and Mallows (1983) to use the term "masking variables" for the irrelevant variables. We denote the masking variables as $V_M = V \setminus V^T$. Masking variables are problematic for applied analyses because they can prevent recovery of true structure in the data and, subsequently, yield erroneous conclusions.

The problem of masking variables is well known in the psychometric and statistical literature for cluster analysis (Brusco & Cradit, 2001; Carmone, Kara, & Maxwell, 1999; Fowlkes, Gnanadesikan, & Kettenring, 1988; Gnanadesikan, Kettenring, & Tsao, 1995). Not surprisingly, a significant research effort has been devoted to the selection of variables for a cluster analysis, as well as differential weighting of variables (Bishop, 1995; Brusco, 2004; Brusco & Cradit, 2001; Carmone et al., 1999; DeSarbo, Carroll, Clark, & Green, 1984; Fowlkes et al., 1988; Friedman & Meulman, 2004; Gnanadesikan et al., 1995; Green, Carmone, & Kim, 1990; Milligan, 1989). Gnanadesikan et al. (1995) observed that variable-weighting procedures, such as those tested by DeSarbo et al. (1984) and Milligan (1989), were frequently outperformed by a variable selection method proposed by Fowlkes et al. (1988). In addition to superior performance, variable selection methods have another important advantage over variable-weighting procedures. Specifically, variable selection procedures exclude masking variables completely, which precludes the need for their measurement in subsequent cluster analyses.

The primary limitation of Fowlkes et al.'s (1988) variable selection procedure is that it requires informal interpretation of graphical information and is, therefore, not conducive to large-scale experimental analyses. Carmone et al. (1999) and Brusco and Cradit (2001) independently developed variable selection procedures that utilize Hubert and Arabie's (1985) adjusted Rand index, which is a measure of agreement between two partitions. This index yields a value of one for perfect agreement, whereas values near zero are indicative of near-chance agreement. The adjusted Rand index has been identified as the most effective external criterion for cluster validation (Milligan & Cooper, 1986). The variable selection procedures developed by Carmone et al. and Brusco and Cradit have proven quite successful at eliminating masking variables; however, they both operate iteratively and are not guaranteed to select an optimal subset.

13.2 A Branch-and-Bound Approach to Variable Selection

The literature base in feature selection for pattern recognition reveals a somewhat different approach than the variable selection heuristics described by Brusco and Cradit (2001) and Carmone et al. (1999). The objective is to select a subset of variables that minimizes (or maximizes) an appropriately selected criterion, such as (12.1). Unfortunately, as we have seen in previous chapters, the appropriate objective criterion in a cluster analysis can, in and of itself, pose a difficult combinatorial optimization problem. Thus, if one is to implement a branch-and-bound procedure for variable selection in cluster analysis, the computational plausibility of such a strategy can be impaired by the particular clustering criterion under consideration. Nevertheless, we shall persevere with an integration of variable selection and cluster analysis.

To illustrate the branch-and-bound procedure for variable selection, we return to equation (2.3) from cluster analysis. Given a set of n multivariate observations in D-dimensional space and the construction of a symmetric dissimilarity matrix, $\mathbf{A}$, as consisting of squared Euclidean distances between pairs of objects, a solution of (2.3) minimizes the within-cluster sums of squared deviations of the objects from their cluster centroids for some number of clusters, K. Now, suppose that the prob-

13.2 A Branch-and-Bound Approach to Variable Selection

lem is extended to the selection of exactly d ($0 < d < D$) variables such that (2.3) is minimized across all $\binom{D}{d} = \left(\frac{D!}{d!(D-d)!}\right)$ possible subsets of size d from D variables. Denoting V_d as a subset of d selected variables, the problem can be represented using the following mathematical programming formulation:

$$\min_{V_d \subset V} \left[\min_{\pi_K \in \Pi_K} \left[\sum_{k=1}^{K} \left[\frac{\sum_{(i<j) \in S_k} a_{ij}(V_d)}{N_k} \right] \right] \right], \quad (13.1)$$

subject to: $\quad |V_d| = d, \quad (13.2)$

where $\quad a_{ij}(V_d) = \sum_{v \in V_d} (x_{iv} - x_{jv})^2, \quad$ for $1 \le i \ne j \le n$. $\quad (13.3)$

The objective function (13.1) represents the minimization of (2.3) across all possible subsets of size d. Constraint (13.2) guarantees that exactly d objects are selected, and equation (13.3) indicates that the dissimilarity matrix $\mathbf{A}(V_d)$ is defined for a particular subset of d variables under consideration. Using an exhaustive enumeration approach, the dissimilarity matrix $\mathbf{A}(V_d)$ would be produced for each possible subset. For each matrix instance, the branch-and-bound algorithm illustrated in Chapter 5 could be applied to obtain the optimal solution for (2.3). The minimum values of $f_3^*(\pi_K)$ across all subsets would then be used to determine the selected subset. Exhaustive enumeration of all subsets in this manner might be possible for modestly sized values of D and d; however, this becomes computational infeasible very quickly. For example, the number of possible subsets of size $d = 10$ from $D = 30$ candidate variables is approximately 30 million.

The branch-and-bound methodology builds subsets of size $d \le D$. An effective branch-and-bound strategy should eliminate variable subsets based on excessive values of (13.1) before reaching subsets of size d. Elimination of a subset of size $d_1 < d$, precludes the need to evaluate all complete subsets of size d that could be realized from augmentation of the subset of size d_1. For example, if $D = 30$ and $d = 10$, the elimination of most subsets with five or fewer variables would be very beneficial, because the total number of subsets for $1 \le d_1 \le 5$ is only 174,436, which is much less than 30 million. Our branch-and-bound algorithm for variable selection uses the vector $\boldsymbol{\gamma} = [\gamma_1, \gamma_2, ..., \gamma_p]$ as representing selected variable indices for a partial solution with p variables. Specifically, γ_j is the variable index in position p of the selected subset ($1 \leq j \leq p$). Further, $\boldsymbol{\gamma}^*$ denotes the incumbent (best found) complete subset of d variables, and $f(\boldsymbol{\gamma}^*)$ is the incumbent criterion value. The steps of the algorithm are as follows:

Step 0. INITIALIZE. Establish an upper bound for the criterion value $f(\boldsymbol{\gamma}^*)$ using a efficient heuristic method. Set $p = 0$, $\gamma_j = 0$ for $j = 1,..., d$.

Step 1. BRANCH FORWARD. Set $p = p + 1$. If $p = 1$ then set $\gamma_p = 1$, otherwise $\gamma_p = \gamma_{p-1} + 1$.

Step 2. FEASIBILE COMPLETION TEST. If $D - \gamma_p < d - p$, then go to Step 5.

Step 3. PARTIAL SOLUTION EVALUATION. If passed, then go to Step 4. Otherwise, go to Step 5.

Step 4. COMPLETE SOLUTION? If $p \neq d$, then go to Step 1. Otherwise, set $\boldsymbol{\gamma}^* = \boldsymbol{\gamma}$ and compute $f(\boldsymbol{\gamma}^*)$.

Step 5. DISPENSATION. If $\gamma_p < D$, then go to Step 6, otherwise go to Step 7.

Step 6. BRANCH RIGHT. Set $\gamma_p = \gamma_p + 1$, and return to Step 2.

Step 7. RETRACTION. Set $\gamma_p = 0$, $p = p - 1$. If $p > 0$, then go to Step 5. Otherwise, return the incumbent solution, which is an optimal solution, and STOP.

An initial lower bound for the criterion index can be obtained using a variable selection heuristic, such as those described by Carmone et al. (1999) or Brusco and Cradit (2001). Alternatively, a random selection of d variables could be used to produce an initial subset. A simple replacement heuristic could be applied to improve the initial selection. Specifically, each of the d selected variables would be considered for replacement by each of the unselected variables. If a replacement improved the value of the within-cluster sums of squares, then the unselected variable would replace the selected variable in the selected subset. This process would be repeated until no improvement in the within-cluster sums of

13.2 A Branch-and-Bound Approach to Variable Selection

squares could be realized by exchanging one of the selected variables for an unselected variable.

Step 1 of the branch-and-bound algorithm advances the pointer and assigns a variable to position p of the selected subset. If the feasibility test in Step 2 fails, then there are not enough objects remaining to complete a subset of size d. Partial solution evaluation in Step 3 is implemented by using the branch-and-bound algorithm for (2.3) as described in Chapter 5 to obtain an optimal partition in K clusters for a matrix that is produced using the pairwise Euclidean distances for objects based on the selected set of variables. If the optimal criterion value is less than the current bound, $f(\gamma^*)$, then processing moves to Step 4; otherwise, the partial subset of variables cannot possibly produce a solution that is better than the incumbent bound, and control passes to Step 5 for dispensation of the partial solution.

If Step 4 is reached and $p = d$, then a variable subset of size d that produces a new best found criterion value has been obtained. The corresponding subset is stored as the new incumbent by setting $\gamma^* = \gamma$, and the complete solution is passed to Step 5 for dispensation. If $p < d$ at Step 4, then a completed subset of size d has not yet been found, and processing returns to Step 1 to move deeper into the tree. Note that in Step 1, the next position in the subset is assigned the first variable index that is larger than the index for the previously assigned position. Thus, for any partial subset, the variable indices in the γ array are in ascending order.

At Step 5, a partial solution is passed to Step 6 (branch right) if the current position can be assigned a higher-indexed variable. However, if variable v_D is selected for the current position (i.e., if $\gamma_p = D$), then processing moves to Step 7 for retraction. Branching right in Step 6 involves nothing more than incrementing the variable index for position p and returning to Step 2. Retraction requires backtracking in the set of assigned objects. If the backtracking results in $p = 0$, then the algorithm terminates. Otherwise, the partial solution is passed back to Step 5 for dispensation.

The branch-and-bound algorithm for variable selection is somewhat simpler than the algorithm described in Chapter 2 for cluster analysis because there is no assignment of objects to positions. The information keeping only requires collection of selected variables, not cluster assignment to variables. Nevertheless, the evaluation step of the variable selection algorithm can require the implementation of the cluster analysis branch-and-bound algorithm for each subset, which can greatly increase computation time.

13.3 A Numerical Example

To demonstrate the branch-and-bound algorithm for variable selection, we use the data reported in Table 13.1. The data consist of Likert-scale measurements for $n = 20$ objects on $D = 6$ variables. The objective is to select exactly $d = 3$ of the $D = 6$ variables so as to identify the subset that yields a minimum value for (2.3). Because there are only $6! / (3!3!) = 20$ feasible subsets of size 3, the solution of this problem via exhaustive enumeration would be trivial. Thus, we must recognize that the purpose of the demonstration is solely to illustrate the branch-and-bound variable selection procedure.

Table 13.1. Example data (objects measured on $D = 6$ Likert-scale variables).

Object	v_1	v_2	v_3	v_4	v_5	v_6
1	2	2	3	4	4	2
2	7	2	2	6	1	7
3	2	3	6	5	5	5
4	1	5	5	7	1	6
5	7	3	6	1	6	2
6	6	5	7	7	6	3
7	7	5	6	4	7	2
8	2	5	5	6	1	5
9	7	4	2	7	2	6
10	1	7	3	7	5	2
11	2	5	2	2	5	3
12	6	5	6	6	7	3
13	2	2	6	1	4	5
14	6	7	2	2	4	7
15	1	5	3	4	7	2
16	6	4	1	4	1	7
17	7	1	7	3	5	3
18	2	2	6	2	6	5
19	7	5	1	2	5	6
20	2	5	3	2	2	3

Table 13.2 presents the results of the application of the branch-and-bound procedure for variable selection to the data in Table 13.1 under the assumption of $K = 4$ clusters. An initial lower bound of 13.3 obtained by a variable selection heuristic was assumed. The results in Table 13.2 indicate that many of the two-variable solutions produced objective-function values that exceeded the lower bound, thus obviating the need to analyze the three-cluster solution. Among these two variable solutions are $\{v_1, v_2\}$, $\{v_1, v_4\}$, $\{v_1, v_5\}$, $\{v_2, v_3\}$, $\{v_2, v_4\}$, $\{v_2, v_5\}$, $\{v_3, v_4\}$, $\{v_3, v_5\}$,

{v_4, v_5}. Although the importance of this pruning loses some of its luster because the size of this problem is small, the fundamental aspect of the approach is evident. That is, for problems of practical size, the branch-and-bound approach enables variable subsets to be implicitly evaluated and eliminated before they need to be explicitly evaluated.

Table 13.2. Variable selection using criterion (2.3) on the data in Table 13.1.

Row	Variables	p	Criterion	Dispensation
1	{v_1}	1	0.00000	Branch forward
2	{v_1, v_2}	2	15.29762	Prune (15.3 ≥ 13.3), Branch right
3	{v_1, v_3}	2	8.80000	Branch forward
4	{v_1, v_3, v_4}	3	80.06667	Suboptimal, 80.07 ≥ 13.3, Branch right
5	{v_1, v_3, v_5}	3	51.94286	Suboptimal, 51.94 ≥ 13.3, Branch right
6	{v_1, v_3, v_6}	3	13.20000	*New Incumbent, $f^* = 13.2$
7	{v_1, v_4}	2	22.41667	Prune (22.42 ≥ 13.2), Branch right
8	{v_1, v_5}	2	20.09524	Prune (20.10 ≥ 13.2), Retract
9	{v_2}	1	1.80000	Branch forward
10	{v_2, v_3}	2	21.06667	Prune (21.07 ≥ 13.2), Branch right
11	{v_2, v_4}	2	25.08333	Prune (25.08 ≥ 13.2), Branch right
12	{v_2, v_5}	2	20.33333	Prune (20.33 ≥ 13.2), Retract
13	{v_3}	1	2.83333	Branch forward
14	{v_3, v_4}	2	26.21429	Prune (26.21 ≥ 13.2), Branch right
15	{v_3, v_5}	2	21.33333	Prune (21.33 ≥ 13.2), Retract
16	{v_4}	1	2.97857	Branch forward
17	{v_4, v_5}	2	30.27381	Prune (30.27 ≥ 13.2), Retract
18	{v_5}	1		TERMINATE

The optimal solution for the variable selection problem is identified in row 6 of Table 13.2, indicating that the optimal subset of variables is {v_1, v_3, v_6}. The optimal partition of objects corresponding to this optimal subset is shown in Table 13.3. Each of the four clusters in Table 13.3 contains five objects, and the similarity of within-cluster measurements across the three selected variables clearly reveals considerable cluster homogeneity. The table also reveals the lack of within-cluster homogeneity with respect to the unselected variables {v_2, v_4, v_5}.

13.4. Strengths, Limitations, and Extensions

The branch-and-bound procedure described above is very interesting because one branch-and-bound routine is embedded within another branch-and-bound routine. In other words, the branch-and-bound approach from

Chapter 5 is incorporated in the algorithm for variable selection, effectively serving as the "engine" that generates the solutions. As described, the approach is *comprehensively* optimal for selecting subsets of size d from D candidates because the solutions for the evaluation for each subset are guaranteed optimal, and the variable selection branch-and-bound approach ensures that all feasible subsets are either explicitly or implicitly evaluated.

Table 13.3. Cluster assignments for optimal solution to the variable selection problem.

	Object	v_1	v_3	v_6	v_2	v_4	v_5
Cluster 1	9	7	2	6	4	7	2
	16	6	1	7	4	4	1
	2	7	2	7	2	6	1
	14	6	2	7	7	2	4
	19	7	1	6	5	2	5
Cluster 2	3	2	6	5	3	5	5
	4	1	5	6	5	7	1
	18	2	6	5	2	2	6
	8	2	5	5	5	6	1
	13	2	6	5	2	1	4
Cluster 3	5	7	6	2	3	1	6
	17	7	7	3	1	3	5
	12	6	6	3	5	6	7
	6	6	7	3	5	7	6
	7	7	6	2	5	4	7
Cluster 4	10	1	3	2	7	7	5
	11	2	2	3	5	2	5
	1	2	3	2	2	4	4
	15	1	3	2	5	4	7
	20	2	3	3	5	2	2

If n and K were sufficiently large so as to preclude successful implementation of the branch-and-bound algorithm for (2.3), the variable selection branch-and-bound procedure could employ the K-means heuristic procedure (MacQueen, 1967) for the solution evaluation process. The algorithm would no longer produce a guaranteed, comprehensively optimal solution; however, tremendous improvements in efficiency would be realized.

In addition to computational feasibility issues associated with problem size, there are at least two important limitations to the approach described in this chapter. First, for the approach to be of pragmatic value, the candidate variables should be measured on approximately the same scale. In some situations, this might not present a problem. For example, if all candidate variables were binary, or all variables were measured on a 7-point Likert scale, then there might be sound justification for this type of approach. However, if the variables were measured on widely different scales, variables of greater magnitude might be unfairly penalized in the selection process. Of course, one approach in such cases would be to standardize the variables; however, variable standardization in cluster analysis is a rather thorny issue (see, for example, Milligan, 1996).

A second drawback is that, although the algorithm described in this chapter provides an optimal solution for a fixed subset size d, the problem of choosing among different values of d remains. Furthermore, in most practical implementations, the problem would be further compounded by the fact that K is also typically unknown. Thus, the analyst is faced with the problem of simultaneously selecting an appropriate subset size, d, and an appropriate number of clusters, K, and clearly the two decisions are not independent of one another. Nevertheless, the algorithm can be implemented for various values of d and K. A selection of the appropriate size can be made using percentage changes in the clustering indices (as described in Chapter 3).

We have described the variable selection problem in cluster analysis with respect to criterion (2.3); however, the general principles are adaptable to a wide spectrum of criteria. For example, the algorithm could be applied using (2.1) or (2.2). Hand (1981b, Chapter 6) and Bishop (1995) describe variable selection problems within the context of misclassification rates as the relevant objective criteria. Any criterion that possesses the monotonicity property could be considered.

14 Variable Selection for Regression Analysis

14.1 Regression Analysis

In the preceding chapter, we described the application of branch-and-bound methods for the selection of variables for cluster analysis and pattern recognition. There are, however, other important variable selection applications and most notable among these is the problem of identifying an appropriate subset of variables for a *regression analysis*. Rather than selecting a subset of variables so as to partition a data set into homogeneous clusters, the variable-selection emphasis in regression analysis is to produce a subset of variables from a set of D candidate *independent* variables ($V = \{v_1, v_2, ..., v_D\}$) that facilitates prediction of a single *dependent* variable (y).

Given n objects, dependent variable measurements y_i ($i = 1,..., n$), and independent variable measurements x_{ij} ($i = 1,..., n$ and $j = 1,..., D$), the predictive linear model for the full set of candidate independent variables is:

$$\hat{y}_i = b_0 + b_1 x_{i1} + b_2 x_{i2} + ... + b_D x_{iD}, \tag{14.1}$$

where b_0 is an intercept term and b_j ($j = 1,..., D$) is the slope coefficient for independent variable v_j.

The most common criterion in regression analysis for choosing the intercept and slope coefficients is the residual (or error) sum of squares:

$$RSS_y = \sum_{i=1}^{n}(y_i - \hat{y}_i)^2. \tag{14.2}$$

If we define $\mathbf{y}$ as the $n \times 1$ vector of dependent variable measurements, $\mathbf{X}$ as an $n \times (D + 1)$ matrix containing 1s in the first column and the independent variable measurements for variables $j = 1,..., D$ in columns 2,..., $D+1$, respectively, and $\mathbf{b}$ as a $(D + 1) \times 1$ vector of variables associated with the slope and intercept terms of (14.1), then the Residual Sum of Squares (RSS_y) can be written in matrix form as:

$$RSS_y = (\mathbf{y} - \mathbf{Xb})' (\mathbf{y} - \mathbf{Xb}). \qquad (14.3)$$

Using matrix differentiation and solving the resulting first-order conditions, the optimal solution to (14.3) is established as:

$$\mathbf{b} = (\mathbf{X'X})^{-1}\mathbf{X'y}. \qquad (14.4)$$

A common measurement of model fit in regression analysis is the coefficient of determination, R^2, which represents the proportion of variation (total sum of squares) in the dependent variable that is explained by the independent variables. The R^2 value is computed as follows:

$$R^2 = (TSS_y - RSS_y) / TSS_y, \qquad (14.5)$$

$$\text{where} \quad TSS_y = \sum_{i=1}^{n}(y_i - \bar{y})^2, \qquad (14.6)$$

$$\text{and} \quad \bar{y} = \frac{1}{n}\sum_{i=1}^{n} y_i. \qquad (14.7)$$

The solution of (14.4) minimizes RSS_y and maximizes R^2. An important aspect of these criteria is that they exhibit the monotonicity property described in Chapter 12. Specifically, the deletion of a variable from the selected set of variables cannot possibly reduce the residual sum of squares or increase R^2. Similarly, when an independent variable is added to the model, the residual sum of squares cannot increase, i.e., we cannot possibly add a variable to the predictive model and explain less variation in the dependent variable.

The reader should note an important difference between the variable selection problem in cluster analysis described in Chapter 13 and the variable selection problem in regression. In the cluster analysis context, the selected criterion (2.3) worsens as variables are *added* to the selected subset. That is, the within-cluster sum of squared deviations increases as variables are added. In the regression context, however, the RSS_y and R^2 values worsen as independent variables are *deleted* from the subset. To reiterate, removing a predictor from the model can only damage the explanation of variation in the dependent variable.

14.1 Regression Analysis

Fortunately, the fact that removal of independent variables from the selected subset worsens RSS_y does not pose serious difficulties for the design of a branch-and-bound algorithm. The same paradigm described in Chapter 13 can be used for variable selection in regression, with only a few minor modifications. First, d is defined as the number of variables to be *eliminated* from D variables—i.e., the desired subset size is $D - d$ independent variables—and $\gamma = [\gamma_1, \gamma_2, ..., \gamma_p,]$ will contain those variables that have been removed from the complete set of selected variables, as opposed to the selected set of variables. For example, if $D = 10$, $p = 3$, and $\gamma = [3, 5, 6]$, then the evaluation step of the algorithm will assess a regression model using variables $\{v_1, v_2, v_4, v_7, v_8, v_9, v_{10}\}$ as predictors for y. When a complete solution with $p = d$ omitted variables is accepted as the new incumbent solution in Step 4, γ^* corresponds to the omitted variables, not the selected variables. The term $f(\gamma^*)$ denotes the incumbent (best found) value of RSS_y across subsets of $D - d$ variables.

These minor modifications notwithstanding, the variable selection algorithm for regression is analogous to the algorithm for cluster analysis. One of the big advantages for the regression context, however, is that the evaluation step (Step 4) does not require the solution of an NP-hard optimization problem. Whereas the variable selection algorithm for cluster analysis required the embedding of a branch-and-bound algorithm for (2.3) in Step 3, the regression algorithm only requires the matrix solution for (14.4), which is much easier to implement.

In summary, the variable selection problem for regression analysis is posed as the selection of $(D - d)$ of D candidate variables. The objective is to select the subset that minimizes RSS_y (or, equivalently, maximizes R^2). Defining V_d as the subset of d unselected variables and $\overline{V}_d$ as the complement of V_d, which corresponds to the subset of $D - d$ selected variables, the problem can be represented using the following mathematical programming formulation:

$$\min_{V_d \subset V_D} \sum_{i=1}^{n} \left[y_i - b_0 - \sum_{j \in \overline{V}_d} b_j x_{ij} \right]^2, \qquad (14.8)$$

$$\text{subject to:} \qquad |V_d| = d. \qquad (14.9)$$

The solution of this mathematical program using a branch-and-bound algorithm is consistent with the approach described in section 13.2, subject to two important modifications: (a) the pointer indexes positions for variables that are eliminated rather than selected, and (b) the partial solution evaluation requires the solution of the normal equation (14.4), rather than a combinatorial clustering problem. For this reason, the branch-and-bound algorithm for variable selection in regression tends to have greater scalability than the variable selection algorithm for K-means clustering.

14.2 A Numerical Example

To illustrate the branch-and-bound algorithm for variable selection in regression analysis, we consider the data in Table 14.1. These data consist of $D = 6$ candidate independent variables, of which $d = 3$ are to be eliminated and $D - d = 3$ are to be selected for predicting the dependent variable y. The objective is to choose the three variables so as to minimize the residual sum of squares (or, analogously, to maximize R^2) across all subsets of size $D - d$.

Using all $D = 6$ independent variables to predict y yields $RSS_y = 31.97$ and $R^2 = .96679$. This solution represents a best case scenario for a subset of $D - d = 3$ variables. An initial lower bound was obtained by eliminating variables v_4, v_5, and v_6 (i.e., selecting variables v_1, v_2, and v_3). This solution produced an initial upper bound for RSS_y of $f(\gamma^*) = 227.71$. The branch-and-bound solution for the data in Table 14.1 is presented in Table 14.2.

The first row of Table 14.2 provides an efficient pruning of partial solutions, as it reveals that variable v_1 must be among the selected clustering variables (or, equivalently, cannot be among the eliminated variables) for the regression analysis. The removal of this independent variable destroys the explanation of the dependent variable, even when all five of the other independent variables are used. The algorithm proceeds with branching based on variable v_2, finding a new incumbent solution in row 5 (this solution is ultimately proven to be optimal). Later in the table, in row 9, we observe that the property for v_1 also holds for variable v_3. That is, removal of v_3 is too costly with respect to deterioration of the criterion value.

The elimination of the $d = 3$ variables $\{v_2, v_4, v_5\}$, which corresponds the selection of $D - d = 3$ variables $\{v_1, v_3, v_6\}$, produces $RSS_y = 35.36$ and $R^2 = .96326$. Thus, variables v_1, v_3, and v_6 provide nearly the same explanation of variation that is realized when all six variables are included in the predictive model. The pragmatic value of such a solution is a more parsimonious model with nearly the same level of explained variation.

Table 14.1. Example data set for 20 objects measured on $D = 6$ Likert-scale variables.

Object	Independent variables (predictors)						Dependent variable
	v_1	v_2	v_3	v_4	v_5	v_6	y
1	4	4	1	7	5	7	22
2	6	5	2	7	6	6	28
3	2	7	2	5	6	4	17
4	7	3	7	3	6	4	38
5	5	6	6	1	3	7	36
6	1	3	6	2	7	4	20
7	2	2	3	7	7	7	21
8	4	4	1	4	4	3	18
9	5	3	6	1	4	6	35
10	7	6	1	3	7	5	29
11	4	3	3	7	7	5	23
12	4	4	7	7	5	1	24
13	6	3	7	7	6	5	38
14	1	3	2	2	6	7	16
15	6	6	4	5	2	3	30
16	1	7	5	3	4	5	21
17	4	6	3	5	5	6	26
18	3	5	5	3	4	6	29
19	6	7	5	6	1	2	27
20	6	4	5	4	5	1	28
21	6	4	3	7	7	7	33
22	2	1	4	7	5	5	23
23	4	3	3	4	7	7	29
24	2	6	7	6	7	3	23
25	6	3	1	1	6	2	22

Table 14.2. Branch-and-bound summary for variable selection in regression using the data in Table 14.1.

Row	Eliminated variables	p	RSS_y	R^2	Dispensation
1	$\{v_1\}$	1	635.26	.34003	Prune (635.26 ≥ 227.71), Branch right
2	$\{v_2\}$	1	32.01	.96674	Branch forward
3	$\{v_2, v_3\}$	2	371.05	.61452	Prune (371.05 ≥ 227.71), Branch right
4	$\{v_2, v_4\}$	2	34.89	.96375	Branch forward
5	$\{v_2, v_4, v_5\}$	3	35.36	.96326	*New Incumbent, $f(\gamma^*) =$ 35.36
6	$\{v_2, v_4, v_6\}$	3	228.77	.76233	Suboptimal, (228.77 ≥ 35.36), Retract
7	$\{v_2, v_5\}$	2	32.18	.96657	Branch forward
8	$\{v_2, v_5, v_6\}$	3	228.29	.76283	Suboptimal, (228.29 ≥ 35.36), Retract
9	$\{v_3\}$	1	365.97	.61979	Prune (365.97 ≥ 35.36), Branch right
10	$\{v_4\}$	1	34.86	.96379	Branch forward
11	$\{v_4, v_5\}$	2	35.35	.96328	Branch forward
12	$\{v_4, v_5, v_6\}$	3	227.71	.76343	Suboptimal, (227.71 ≥ 35.36), Retract
13	$\{v_5\}$	1			TERMINATE

Termination of the algorithm occurs in row 13. The optimal regression model corresponding to the three selected variables is:

$$\hat{y}_i = .224 + 2.671 x_{i1} + 1.902 b_3 x_{i3} + 1.562 b_6 x_{i6}, \qquad (14.10)$$

which was close to what might be expected, as the data in Table 14.1 were generated using the following equation in conjunction with a small random error component, ε_i.

$$y_i = 3.0 x_{i1} + 2.0 b_3 x_{i3} + 1.5 b_6 x_{i6} + \varepsilon_i. \qquad (14.11)$$

The performance of the branch-and-bound algorithm is sensitive to the ordering of the variables. To illustrate, suppose that objects are considered for elimination in the order $v_1, v_3, v_6, v_2, v_4, v_5$ (we will not renumber the objects to avoid confusion). Table 14.3 provides the results for the branch-and-bound algorithm when using this order. The first three rows of this table reveal effective pruning because they are based on only $p = 1$ variable being eliminated from the set of independent variables (i.e., the tree is pruned at a very early stage). The total number of rows in Table 14.3 is only 10, which compares favorably to the 13 rows in Table 14.2.

Table 14.3. Branch-and-bound summary for variable selection in regression using the data in Table 14.1, assuming variable reordering.

Row	Eliminated variables	p	RSS_y	R^2	Dispensation
1	$\{v_1\}$	1	635.26	.34003	Prune (635.26 ≥ 227.71), Branch right
2	$\{v_3\}$	1	365.97	.61979	Prune (365.97 ≥ 227.71), Branch right
3	$\{v_6\}$	1	220.65	.87793	Branch forward
4	$\{v_6, v_2\}$	2	222.98	.87655	Branch forward
5	$\{v_6, v_2, v_4\}$	3	228.77	.87312	Suboptimal (228.77 ≥ 227.71), Branch right
6	$\{v_6, v_2, v_5\}$	3	228.29	.87340	Suboptimal (228.29 ≥ 227.71), Retract
7	$\{v_2\}$	1	32.01	.96674	Branch forward
8	$\{v_2, v_4\}$	2	34.89	.96375	Branch forward
9	$\{v_2, v_4, v_5\}$	3	35.36	.96326	*New Incumbent, $f^* = 35.36$
10	$\{v_4\}$	1			TERMINATE

14.3 Application to a Larger Data Set

To demonstrate the variable selection problem in regression from a more practical standpoint, we consider a data set corresponding to 21 body measurements and four other variables (age, gender, height, and weight) for $n = 507$ men and women. These data, which were collected by Grete Heinz and Louis J. Peterson at a number of California fitness clubs, were published online in the *Journal of Statistics Education*, by Heinz, Peterson, Johnson, and Kerk (2003). This data set provides an effective tool for teaching multiple regression, as well as other multivariate statistical techniques, such as discriminant analysis. Our focus in this current illustration, of course, is to demonstrate variable selection issues.

The dependent variable in our analysis is subject weight (measured in kilograms), which is consistent with the original description by Heinz et al. (2003). Each of the remaining 24 measurements are considered as candidates for predictor variables. The $D = 24$ candidate variables are displayed in Table 14.4, and are subdivided into three categories: (a) skeletal measurements, (b) girth measurements, and (c) other measurements. All measurements of skeletal structure and girth, as well as height, are in centimeters (cm). Age is measured in years, and gender is a binary variable where 0 = female and 1 = male.

The branch-and-bound-algorithm was applied to the body measurements data set from Heinz et al. (2003) for subset sizes ranging from $1 \leq D - d \leq 15$. As we observed in the example in the previous subsection, the ordering of the variables affects the efficiency of the algorithm. Our Fortran implementation of the algorithm, described in more detail in section 14.5, is initiated with a two-stage heuristic procedure that produces both a strong initial upper bound for RSS_y, as well as information that enables a favorable reordering of objects. The first stage of the heuristic uses a *forward selection* process that adds variables, one at time, based on greatest improvement in RSS_y. An *exchange algorithm* similar to those used in Chapters 3-5 is used in the second stage to see if selected variables can be replaced with unselected variables to improve RSS_y. The rationale of object reordering is that placement of variables with good explanatory power early in the variable list should promote pruning branches early in the tree. Upon completion of the branch-and-bound process, the program translates the variables back to their natural order. The RSS_y values, R^2 values, and corresponding optimal subsets for each subset size are shown in Table 14.5.

An examination of Table 14.5 reveals sharp reductions in the residual sum of squares criterion as the subset size is gradually increased from 1 to about 7. An interesting note is that v_{12} provides the minimum RSS_y value across all candidate predictors; however, if a pair of variables are to be selected based on the RSS_y criterion, then this variable is not included in the optimal pair, which is $\{v_{11}, v_{18}\}$. Variable v_{12} returns to the optimal subset for $D - d = 3$, and remains in the optimal subset at least through $D - d = 15$. Height (v_{23}) also enters the optimal mix at $D - d = 3$ and remains there for the remainder of the table. The first skeletal measurement to become a member of the optimal subset is v_8 at $D - d = 6$. Although this variable briefly leaves the optimal mix at $D - d = 7$, it quickly returns at $D - d = 8$ and remains in the optimal subset throughout the rest of the table. The age (v_{22}) and gender (v_{24}) variables are late entries into the selected subset, joining the optimal mix at $D - d = 9$ and $D - d = 12$, respectively.

Table 14.4. Candidate independent variables from Heinz et al. (2003). The dependent variable is weight (in kilograms).

Independent variable class	Label	Description
Skeletal Measurements (cm)	v_1	Biacromial diameter
	v_2	Biiliac diameter, or "pelvic breadth"
	v_3	Bitrochanteric diameter
	v_4	Chest depth between spine and sternum
	v_5	Chest diameter at nipple level, mid-expiration
	v_6	Elbow diameter, sum of two elbows
	v_7	Wrist diameter, sum of two wrists
	v_8	Knee diameter, sum of two knees
	v_9	Ankle diameter, sum of two ankles
Girth Measurements (cm)	v_{10}	Shoulder girth over deltoid muscles
	v_{11}	Chest girth, mid-expiration
	v_{12}	Waist girth, narrowest part of torso
	v_{13}	Navel (or "Abdominal") girth at umbilicus
	v_{14}	Hip girth at level of bitrochanteric diameter
	v_{15}	Thigh girth below gluteal fold, average of two thighs
	v_{16}	Bicep girth, flexed, average of right and left girths
	v_{17}	Forearm girth, extended, palm up, average of two forearms
	v_{18}	Knee girth over patella, flexed, average of two knees
	v_{19}	Calf maximum girth, average of right and left girths
	v_{20}	Ankle minimum girth, average of right and left girths
	v_{21}	Wrist minimum girth, average of right and left girths
Other Measurements	v_{22}	Age (years)
	v_{23}	Height (cm.)
	v_{24}	Gender (0, 1)

14 Variable Selection for Regression Analysis

Table 14.5. Optimal subsets for the body measurements data.

subset size $(D-d)$	RSS_y	R^2	Selected subset
1	16474.6	.817199	$\{v_{12}\}$
2	8711.1	.903342	$\{v_{11}, v_{18}\}$
3	5179.0	.942534	$\{v_{12}, v_{15}, v_{23}\}$
4	3375.5	.962545	$\{v_{12}, v_{15}, v_{17}, v_{23}\}$
5	2996.6	.966749	$\{v_{11}, v_{12}, v_{15}, v_{19}, v_{23}\}$
6	2742.0	.969575	$\{v_8, v_{11}, v_{12}, v_{15}, v_{19}, v_{23}\}$
7	2510.9	.972139	$\{v_{11}, v_{12}, v_{14}, v_{15}, v_{17}, v_{19}, v_{23}\}$
8	2410.9	.973249	$\{v_8, v_{11}, v_{12}, v_{14}, v_{15}, v_{17}, v_{19}, v_{23}\}$
9	2339.0	.974046	$\{v_8, v_{11}, v_{12}, v_{14}, v_{15}, v_{17}, v_{19}, v_{22}, v_{23}\}$
10	2283.9	.974657	$\{v_4, v_8, v_{11}, v_{12}, v_{14}, v_{15}, v_{17}, v_{19}, v_{22}, v_{23}\}$
11	2248.7	.975048	$\{v_4, v_8, v_{11}, v_{12}, v_{14}, v_{15}, v_{17}, v_{18}, v_{19}, v_{22}, v_{23}\}$
12	2207.9	.975501	$\{v_4, v_8, v_{10}, v_{11}, v_{12}, v_{14}, v_{15}, v_{17}, v_{19}, v_{22}, v_{23}, v_{24}\}$
13	2181.9	.975789	$\{v_4, v_8, v_{10}, v_{11}, v_{12}, v_{14}, v_{15}, v_{17}, v_{18}, v_{19}, v_{22}, v_{23}, v_{24}\}$
14	2167.6	.975949	$\{v_2, v_4, v_8, v_{10}, v_{11}, v_{12}, v_{14}, v_{15}, v_{17}, v_{18}, v_{19}, v_{22}, v_{23}, v_{24}\}$
15	2155.5	.976083	$\{v_2, v_4, v_5, v_8, v_{10}, v_{11}, v_{12}, v_{14}, v_{15}, v_{17}, v_{18}, v_{19}, v_{22}, v_{23}, v_{24}\}$

Table 14.6 presents the regression coefficients, along with their t-statistics and corresponding p-values for the $D - d = 13$ subset size. This table was obtained using Minitab®, version 12.0 (Minitab® is a trademark of Minitab, Inc.).

Table 14.6. Minitab regression output for the optimal 13-variable subset.

Predictor	Coefficient	Std. Dev.	T-statistic	p-value
Intercept	-121.708	2.5240	-48.22	.000
v_4 (chest depth)	.26497	.06846	3.87	.000
v_8 (knee diameter)	.5572	.12170	4.58	.000
v_{10} (shoulder girth)	.08507	.02812	3.03	.003
v_{11} (chest girth)	.1709	.03389	5.04	.000
v_{12} (waist girth)	.37698	.02513	15.00	.000
v_{14} (hip girth)	.22403	.03844	5.83	.000
v_{15} (thigh girth)	.24641	.04872	5.06	.000
v_{17} (forearm girth)	.56662	.09669	5.86	.000
v_{18} (knee girth)	.17977	.07427	2.42	.016
v_{19} (calf girth)	.35343	.06159	5.74	.000
v_{22} (age)	-.05208	.01176	-4.43	.000
v_{23} (height)	.31403	.01584	19.83	.000
v_{24} (gender)	-1.4278	.48490	-2.94	.003
$RSS_y = 2181.9$, $R^2 = 97.6\%$, Adjusted $R^2 = 97.5\%$				

Table 14.6 shows that all 13 predictors are statistically significant at the .05 level, and the RSS_y and R^2 values are in accordance with the Fortran results in Table 14.5. Among the members of the optimal subset are 8 girth measurements, 2 skeletal measurements, height, age, and gender. The 13-predictor solution in Table 14.6 (model A) can be compared to the 13-predictor solution reported by Heinz et al. (2003), which used all 12 girth measurements and height as predictors (we refer to this model as model B). As would be expected, model A provides a slightly better RSS_y value (2181.9) than model B (2394.2). Specifically, four of the girth-measurement predictors in model B—v_{13} (navel girth), v_{16} (bicep girth), v_{20} (ankle girth), and v_{21} (wrist girth)—were not statistically significant. The replacement of these four girth measurements with two skeletal measurements (v_4 (chest depth), v_8 (knee diameter)) along with age (v_{22}) and gender (v_{24}) reduces RSS_y (increases R^2) and leads to a complete set of 13 significant predictors.

Table 14.7. Minitab results after removal of age and gender as predictors.

Predictor	Coefficient	Std. Dev.	T-statistic	p-value
Intercept	-121.26500	2.34000	-51.83	.000
v_4 (chest depth)	.21318	.06926	3.08	.002
v_8 (knee diameter)	.44350	.12230	3.63	.000
v_{10} (shoulder girth)	.08181	.02817	2.90	.004
v_{11} (chest girth)	.17810	.03457	5.15	.000
v_{12} (waist girth)	.32883	.02324	14.15	.000
v_{14} (hip girth)	.22388	.03673	6.10	.000
v_{15} (thigh girth)	.33886	.04630	7.32	.000
v_{17} (forearm girth)	.48190	.09062	5.32	.000
v_{18} (knee girth)	.21742	.07537	2.88	.004
v_{19} (calf girth)	.35264	.06306	5.59	.000
v_{23} (height)	.31072	.01519	20.46	.000
RSS = 2296.7, R^2 = 97.5%, Adjusted R^2 = 97.4%				

The number of possible model refinements and explorations that can be made with the branch-and-bound model is enormous. Nevertheless, we offer one final demonstration, attempting to provide a compromise between model A and model B. We eliminated the age (v_{22}) and gender (v_{24}) variables from the candidate pool and sought to identify a subset of predictors that would provide an effective combination of girth and skeletal variables (in addition to height). The branch and bound algorithm produced the same subset as shown in Table 14.6, with the exception that age and gender are omitted. The Minitab regression results for this subset (Table 14.7) reveal that the removal of age and gender does not substantially impair the explained variation, and this 11-predictor solution still explains more variation than model B.

14.4 Strengths, Limitations, and Extensions

The branch-and-bound algorithm for variable selection in regression analysis can be used to select a subset of $D - d$ variables that minimizes RSS_y across all subsets of size $D - d$. This is obviously of considerable value because it can facilitate the selection of a good subset of variables for a regression analysis. If there are highly correlated independent variables within the candidate set, V_D, the branch-and-bound algorithm has the propensity to select only one member of a highly correlated subset because selection of two highly correlated variables is not apt to add much explanatory value to the regression model. Thus, the branch-and-bound algorithm can have some benefit with respect to avoiding problems associated with multicollinearity (highly correlated independent variables in a regression model). This well-known problem in regression analysis can sometimes lead to incorrect interpretation of independent variable effects because of distortion of the slope coefficients.

The branch-and-bound algorithm, however, is not a foolproof solution to the problem of multicollinearity. Moreover, branch-and-bound selection of variables as described in this chapter does not necessarily eliminate other possible problems in regression modeling, such as non-normality of error terms, nonconstancy of error variance, and outliers. A complete diagnostic evaluation of the regression results would be in order for a solution produced via the branch-and-bound algorithm. Coverage of diagnostics in regression analysis is not within the scope of this monograph; however, many useful resources address these issues (Draper & Smith, 1981; Neter, Wasserman, & Kutner, 1985; Weisberg, 1985).

As was the case for variable selection in cluster analysis, the problem of identifying the appropriate subset size is problematic in the regression context. In some situations, the deterioration of RSS_y or R^2 that results from increasing the number of eliminated variables from d to $d + 1$ might suggest a clear cutoff point for the number of eliminated variables. For example, if the R^2 values for $d = 4, 5, 6$, and 7 are 93%, 90%, 80%, and 77%, respectively, then $d = 5$ might be selected as appropriate. In other situations, incremental changes might be less substantial and perhaps other methods would be in order. Miller (2002) devotes an entire chapter to stopping rules that can be used to facilitate the determination of subset size. Hastie, Tibshirani, and Friedman (2001, Chapter 3) suggest that the subset size should be jointly determined on the basis of bias and variance, and thoroughly address these two issues in Chapter 7 of their book.

Finally, we note that there are a number of extensions of branch-and-bound methods in regression analysis. For example, Leenen and Van Mechelen (1998) described an application of branch-and-bound within the context of Boolean regression. Armstrong and Frome (1976) developed a branch-and-bound procedure for regression where the slope coefficients for independent variables are constrained to be nonnegative (see also Hand, 1981a). The bounding procedure in this application determines which variables should be driven to the boundary, i.e., forced to have coefficients of zero. Variable selection has also been recognized as an especially important problem in logistic regression (Hosmer, Jovanovic, & Lemeshow, 1989; King, 2003).

14.5 Available Software

We have made available two software programs for identifying a subset of independent variables (of fixed size $D - d$) that will minimize RSS_y across all feasible subsets. These software programs, which can be downloaded from http://www.psiheart.net/quantpsych/monograph.html or http://garnet.acns.fsu.edu/~mbrusco, are consistent with the methodology described in this chapter. The first program, *bestsub.for*, is written in Fortran, and uses a "brute force" regression subroutine that solves the normal equations for candidate subsets. The second program is a MATLAB® *.m file, *bestsub.m* (MATLAB® is a trademark of The MathWorks, Inc.). We selected MATLAB for implementation because this software package has a built-in regression command that easily performs the necessary computations for the branch-and-bound algorithm. The resulting program is, therefore, shorter and conceptually easier to understand than our Fortran implementation for the same procedure.

We have incorporated reordering and bounding procedures in *bestsub.for* and *bestsub.m*, which have improved their computational efficiency for larger data sets. Specifically, the programs begin by using forward selection to identify a solution of the desired subset size. Forward selection begins with the selection of the independent variable that yields the greatest explained variation for the dependent variable. Next, a second independent variable is added such that this variable, when used in conjunction with the first selected variable, provides the greatest improvement in explained variation. This procedure continues until $D - d$

variables have been selected, resulting in an incumbent heuristic solution.

The forward selection process is immediately followed by an exchange algorithm that attempts to modify the incumbent heuristic solution if explained variation can be improved. Each selected variable is sequentially considered as a candidate for removal, and each unselected variable is evaluated as its replacement. If an improvement in explained variation is realized by replacing a selected variable with one of the unselected variables, then this change is made in the incumbent heuristic solution. This process continues until a complete evaluation of all selected variables is made so that no improvement is possible via a replacement. The information from the two-stage heuristic procedure (forward selection followed by the exchange heuristic) serves two purposes. First, the residual sum-of-squares from this solution often provides a reasonably tight upper bound (certainly better than a bound associated with a randomly selected subset). Second, and perhaps more importantly, the heuristic solution can be used to produce an effective reordering of the candidate variables. For example, we choose the first $D - d$ variables in the reordering to be those variables selected by the two-stage process. The last d objects in the reordering are generated based on rank-ordered residual sum of squares values for single-predictor models.

The MATLAB branch-and-bound program for variable selection in regression requires three inputs, xcand, ymeas, and nsel. The matrix xcand is an $n \times D$ matrix where the columns correspond to candidate independent variables. The vector ymeas is an $n \times 1$ vector of dependent variable measurements, and nsel is a scalar constant representing the desired subset size, $D - d$. Of course, the user can use different names for these three pieces of data when implementing the program. The software program is invoked from the command line in MATLAB using the statement:

$$[\text{sel,b,rss,r2,cp}] = bestsub(\text{xcand, ymeas, nsel}). \qquad (14.12)$$

The output measures from the program correspond to the terms in the brackets on the left-side of (14.12). The $(D - d) \times 1$ vector sel contains the selected variables. The first element of the vector b is the intercept term, and the remaining $D - d$ components of the vector are the slope coefficients and match one-to-one with the selected variables in sel. The rss and r2 terms are the optimal residual sum of squares and coefficient of determination, respectively. The cp value is Mallows C_p index (Mallows, 1973), which could be useful as a guideline for comparing subsets of different sizes, although other methods described by Miller (2002) and Hastie et al. (2001) might be preferable for this task.

The Fortran program for variable selection reads a matrix of candidate independent variable measurements and a vector of dependent variable measurements from two separate files, matx and maty, respectively. The user must input the number of objects (n), the number of candidate variables (D), and the desired subset size ($D - d$). The program produces output consisting of the selected variables, the residual sum of squares, R^2, adjusted R^2, Mallows C_p index, and total CPU time. We applied the programs *bestsub.for* to the body measurements data described in section 14.3, for which $D = 24$ and we chose $D - d = 13$. The optimal subset was extracted in less than one CPU second, and the results are analogous to those in Table 14.6.

The input information is as follows:

```
> INPUT NUMBER OF CASES
> 507
> INPUT NUMBER OF CANDIDATE PREDICTORS
> 24
> INPUT SUBSET SIZE
> 13
```

The output information is as follows:

```
INITIAL BOUND          2181.969118085830360
ERROR SUM OF SQUARES         2181.9691

ACTUAL VARIABLES REMOVED ARE
16   5  21   6   7  20   9   1  13   3   2
ACTUAL VARIABLES SELECTED ARE
12  18  23  15  17  11  19  14   8  22   4  10  24

ERROR SUM OF SQUARES         2181.9691
R SQUARED                    0.975789
ADJUSTED R SQUARED           0.975151
MALLOWS CP INDEX                16.91
TOTAL CPU TIME =       0.16
Stop - Program terminated.
```

The scalability of *bestsub.for* and *bestsub.m* is dependent on D, $D - d$, and the inherent relationships in the data. The programs use direct solution of the normal equations, which perhaps limits their efficiency relative to other variable selection methods for regression, such as Furnival and Wilson's (1974) "leaps and bounds" algorithm. Another limitation of our programs is that they only produce a single optimal subset of a fixed size, whereas Furnival and Wilson's (1974) algorithm produces both optimal and several near-optimal solutions for subsets of various sizes.

These limitations of our programs notwithstanding, our experience with the programs is encouraging. Hastie et al. (2001) observe that the practical limit for branch-and-bound procedures similar to the one designed by Furnival and Wilson (1974) is roughly $30 \leq D \leq 40$. We have successfully applied *bestsub.m* and *bestsub.for* to data sets with 30 to 50 predictor variables and several good competing regression models within the set of candidate variables (a condition that tends to increase solution difficulty). Even for the relatively larger problem instances, we are frequently able to produce optimal results in just a few minutes of microcomputer CPU time. We believe that the successful performance of our implementations is associated with the variable reordering based on the two-stage heuristic procedures use to establish an initial regression solution.

Although our results with *bestsub.m* and *bestsub.for* do appear promising, there are a variety of possible enhancements that could be undertaken. For example, greater practical utility would be realized from the implementation of procedures to compare the results associated with different subset sizes.

APPENDIX A

General Branch-and-Bound Algorithm for Partitioning

In the following algorithm, comments are enclosed in slashes and asterisks, "/*" and "*/". Instructions for routines that are not explicit in the pseudocode are enclosed in curly brackets, "{" and "}".

{ **Determine** number of objects, n, and number of clusters, num_k. }
{ **Establish** initial *incumbent* and BestSolution for lambda[n] using an efficient heuristic method. To be used in PARTIAL_EVAL. We can use this initial *incumbent* to find an advantageous reordering of the object list.}
/* **Initialize** incumbents, variables and parameters */
Position = 0   /* Position pointer */
EmptyC = num_k /* Number of empty clusters */
for j = 1 to n /* Cluster assignments of objects */
  lambda(j) = 0
next j
for k = 1 to num_k /* Cluster sizes */
  CSize(k) = 0
next k
Terminate = False
while not Terminate
  /* **Branch Forward** */
  Position = Position + 1
  k = 1
  CSize(k) = CSize(k) + 1
  lambda(Position) = k
  if CSize(k) = 1 then EmptyC = EmptyC − 1
  Dispensation = True
  /* Dispensation indicates that we will Retract and/or Branch Right */
  while Dispensation and not Terminate
    if n − Position >= EmptyC then

/* If we have more empty clusters than available/unselected objects,
then, Dispensation remains True until we backtrack to a feasible
assignment */
Dispensation = not PARTIAL_EVAL
if not Dispensation then
  if *Position* = *n* then
    /* If sequence is complete, then we need to retract */
    Dispensation = True
    /* In PARTIAL_EVAL, we checked to see whether the
sequence could lead to a better solution. But, if the sequence
was actually complete when it passed PART_EVAL, then we
have a better solution */
    incumbent = EVALUATION
    BestSolution(i) = lambda(i) (for i = 1 to *n*)
  end if
  end if
end if
/* **Dispensation** occurs when the current partial solution is not feasible
or will not lead to an optimal solution. Retraction is performed
until we can Branch Right or Terminate the algorithm */
if Dispensation then
  while k = *num_k* or CSize(k) = 1
    /* **Retraction** */
    lambda(*Position*) = 0
    CSize(k) = CSize(k) – 1
    Position = *Position* – 1
    if CSize(k) = 0 then EmptyC = EmptyC + 1
    k = lambda(*Position*)
    if *Position* = 0 then Terminate = True
  loop
  if not Terminate then
    /* **Branch Right** */
    CSize(k) = CSize(k) – 1
    k = k + 1
    CSize(k) = CSize(k) + 1
    if CSize(k) = 1 then EmptyC = EmptyC – 1
    lambda(*Position*) = k
  end if
end if   /* Dispensation, Retraction and Branch Right */
loop   /* Feasibility/Dispensation loop */
loop   /* Termination loop */

APPENDIX B

General Branch-and-Bound Algorithm Using Forward Branching for Optimal Seriation Procedures

This algorithm assumes that the number of objects, n, has been previously determined. Other pertinent variables are for the position pointer (*Position*), the array of n possible positions for the objects (*permutation*), the incumbent solution (*incumbent*), and the array holding the objects in the order of the best solution found at any time during the execution of the algorithm (*BestSolution*). Secondary Boolean variables (*NotRetract, NotRedundancy, found,* and *fathom*) control the flow of the algorithm or assist in determining the final object to be assigned a place in the *permutation*.

Four functions—EVALUATION (Real), ADJACENCYTEST (Boolean) or INTERCHANGETEST (Boolean), and BOUNDTEST (Boolean)—are dependent on the criteria being implemented. The ADJACENCYTEST compares the contribution to the objective function value of swapping *permutation*(*Position* – 1) with *permutation*(*Position*); if the contribution is greater, then the adjacency test fails, we prune the branch and we move to the next branch. The ADJACENCYTEST can be replaced by an INTERCHANGETEST, which extends the adjacency test to compare effects on the objective function value by swapping the candidate for *permutation*(*Position*) with objects previously assigned to positions in *permutation*. The variable *LowerB* is determined in the algorithm, initially set to 0 or calculated by the combinatorial heuristic suggested by Hubert and Arabie (1994). The determination of the initial lower bound parallels the INITIALIZE step in branch-and-bound procedures for cluster analysis. The BOUNDTEST calculates an upper bound given a partial sequence of the n objects; if the upper bound for the partial sequence is less than *LowerB*, then the branch is pruned and we branch right.

/* **Set** lower bound on objective function value */

LowerB = {0 or determined by heuristic}
/* **Initialize** position locator in first position */
Set *Position* = 1 and *permutation*(*Position*) = 1
/* **Initialize** all other positions as "unfilled" */
for k= 2 to *n*
  permutation (k) = 0
next k
/* **Continue** until termination criteria is met.*/
while (*Position* <> 1 or *permutation*(*Position*) <= *n*)
  /* **Increment** position locator */
  Position = *Position* + 1
  fathom = false
  /* **Fathom** next branch */
  while not *fathom*
    /* **Fathom** next branch by incrementing value in "*Position*" of
      permutation */
    permutation(*Position*) = *permutation*(*Position*) + 1
    /* **Check Redundancy** */
    NotRedundancy = True
    for k = 1 to *Position* − 1
      if *permutation*(*Position*) = *permutation*(k) then
        NotRedundancy = False
    next k
    if *NotRedundancy* then
      /* If termination criteria are met, then **exit** loop and algorithm */
      if (*Position* = 1 and *permutation*(*Position*) > *n*) then exit loop
      NotRetract = true
      if (*Position* > 1 and *permutation*(*Position*) > *n*) then
        /* **Retraction** retries previous "*Position*" */
        /* If retracted, then remainder of loop is ignored */
        NotRetract = false
        permutation(*Position*) = 0
        Position = *Position* − 1
      end if
      if *NotRetract* then
        if *Position* = *n* − 1 then
          /* **Find** remaining object for *permutation*(*n*) */
          for i = 1 to *n*
            found = False
            for j = 1 to *n* − 1
              if *permutation*(j) = i then *found* = True

```
            next j
            if not found then permutation(n) = i
          next i
          /* Evaluate when complete sequence is ready */
          incumbent = EVALUATION
          if incumbent > LowerB then
            LowerB = incumbent
            BestSolution = permutation
          end if
        else
          /* Perform fathoming tests. If either test fails, then we
          remain in this loop—incrementing the object in Position
          until permutation(Position) > n */
          if ADJACENCYTEST then fathom = BOUNDTEST
        end if
      end if    /* No Retraction */
    end if    /* No Redundancy */
  loop    /* fathom loop */
loop    /* Termination loop */
/* Return BestSolution as optimal permutation and LowerB as optimal
objective function value */
```

References

Anderberg, M. R. (1973). *Cluster analysis for applications.* New York: Academic Press.

Armstrong, R. D., & Frome, E. L. (1976). A branch and bound solution of a restricted least squares problem. *Technometrics, 18,* 447-450.

Baker, F. B., & Hubert, L. J. (1976). A graph-theoretic approach to goodness of fit in complete-link hierarchical clustering. *Journal of the American Statistical Association, 71,* 870-878.

Banfield, C. F., & Bassil, L. C. (1977). A transfer algorithm for nonhierarchical classification. *Applied Statistics, 26,* 206-210.

Barthélemy, J.-P., Hudry, O., Isaak, G., Roberts, F. S., & Tesman, B. (1995). The reversing number of a digraph. *Discrete Applied Mathematics, 60,* 39-76.

Beale, E. M. L., Kendall, M. G., & Mann, D. W. (1967). The discarding of variables in multivariate analysis. *Biometrika, 54,* 357-366.

Bellman, R. (1962). Dynamic programming treatment of the traveling salesman problem. *Journal of the Association for Computing Machinery, 9,* 61-63.

Bishop, C. M. (1995). *Neural networks for pattern recognition.* New York: Oxford University Press.

Blin, J. M., & Whinston, A. B. (1974). A note on majority rule under transitivity constraints. *Management Science, 20,* 1439-1440.

Bowman, V. J., & Colantoni, C. S. (1973). Majority rule under transitivity constraints. *Management Science, 19,* 1029-1041.

Bowman, V. J., & Colantoni, C. S. (1974). Further comments on majority rule under transitivity constraints. *Management Science, 20,* 1441.

Brown, J. R. (1972). Chromatic scheduling and the chromatic number problem. *Management Science, 19,* 456-463.

Brucker, F. (1978). On the complexity of clustering problems. In M. Beckmann & H. P. Kunzi (Eds.), *Optimization and operations research* (pp. 45-54), Heidelberg: Springer-Verlag.

Brusco, M. J. (2001). Seriation of asymmetric proximity matrices using integer linear programming. *British Journal of Mathematical and Statistical Psychology, 54,* 367-375.

Brusco, M. J. (2002a). Integer programming methods for seriation and unidimensional scaling of proximity matrices: A review and some extensions. *Journal of Classification, 19,* 45-67.

Brusco, M. J. (2002b). A branch-and-bound algorithm for fitting anti-Robinson structures to symmetric dissimilarity matrices. *Psychometrika, 67,* 459-471.

Brusco, M. J. (2002c). Identifying a reordering of the rows and columns of multiple proximity matrices using multiobjective programming. *Journal of Mathematical Psychology, 46*, 731-745.

Brusco, M. J. (2003). An enhanced branch-and-bound algorithm for a partitioning problem. *British Journal of Mathematical and Statistical Psychology, 56*, 83-92.

Brusco, M. J. (2004). Clustering binary data in the presence of masking variables. *Psychological Methods, 9*, 510-523.

Brusco, M. J. (2005). A repetitive branch-and-bound procedure for minimum within-cluster sums of squares partitioning. *Psychometrika*, in press.

Brusco, M. J., & Cradit, J. D. (2001). A variable-selection heuristic for *k*-means clustering. *Psychometrika, 66*, 249-270.

Brusco, M. J., & Cradit, J. D. (2004). Graph coloring, minimum-diameter partitioning, and the analysis of confusion matrices. *Journal of Mathematical Psychology, 48*, 301-309.

Brusco, M. J., & Cradit, J. D. (2005). Bicriterion methods for partitioning dissimilarity matrices. *British Journal of Mathematical and Statistical Psychology*, in press.

Brusco, M. J., Cradit, J. D., & Stahl, S. (2002). A simulated annealing heuristic for a bicriterion partitioning problem in market segmentation. *Journal of Marketing Research, 39*, 99-109.

Brusco, M. J., Cradit, J. D., & Tashchian, A. (2003). Multicriterion clusterwise regression for joint segmentation settings: An application to customer value. *Journal of Marketing Research, 40*, 225-234.

Brusco, M. J., & Stahl, S. (2001a). An interactive approach to multiobjective combinatorial data analysis. *Psychometrika, 66*, 5-24.

Brusco, M. J., & Stahl, S. (2001b). Compact integer programming models for extracting subsets of stimuli from confusion matrices. *Psychometrika, 66*, 405-419.

Brusco, M. J., & Stahl, S. (2004). Optimal least-squares unidimensional scaling: Improved branch-and-bound procedures and comparison to dynamic programming. *Psychometrika*, in press.

Brusco, M. J., & Stahl, S. (2005). Bicriterion methods for skew-symmetric matrices. *British Journal of Mathematical and Statistical Psychology*, in press.

Carmone, F. J., Kara, A., & Maxwell, S. (1999). HINoV: A new model to improve market segmentation by identifying noisy variables. *Journal of Marketing Research, 36*, 501-509.

Charon, I., Guénoche, A., Hudry, O., & Woirgard, F. (1997). New results on the computation of median orders. *Discrete Mathematics, 165/166*, 139-153.

Charon, I., Hudry, O., & Woigard, F. (1996). Ordres médians et ordres de Slater de tournois. *Mathématiques, Informatique, et Sciences Humaines, 34*, 23-56.

Chen, G. (2000). *Metric two-way multidimensional scaling and circular unidimensional scaling: Mixed integer programming approaches.* Doctoral dissertation, Department of Management Science and Information Systems, Rutgers University, Newark, NJ.

Cho, R. Y., Yang, V., & Hallett, P. E. (2000). Reliability and dimensionality of judgments of visually textured materials. *Perception & Psychophysics, 62*, 735-752.

Cobby, J. M. (1986). AS R67: A remark on AS 199, a branch and bound algorithm for determining the optimal feature subset of a given size. *Applied Statistics, 35*, 314.

Cobby, J. M. (1991). Correction to remark AS 67 – a remark on AS 199, a branch and bound algorithm for determining the optimal feature subset of a given size. *Applied Statistics, 40*, 376-377.

DeCani, J. S. (1969). Maximum likelihood paired comparison ranking by linear programming. *Biometrika, 56*, 537-545.

DeCani, J. S. (1972). A branch and bound algorithm for maximum likelihood paired comparison ranking by linear programming. *Biometrika, 59*, 131-135.

Defays, D. (1978). A short note on a method of seriation. *British Journal of Mathematical and Statistical Psychology, 31*, 49-53.

Delattre, M., & Hansen, P. (1980). Bicriterion cluster analysis. *IEEE Transactions on Pattern Analysis and Machine Intelligence, 2*, 277-291.

DeSarbo, W. S., Carroll, J. D., Clark, L. A., & Green, P. E. (1984). Synthesized clustering: A method for amalgamating alternative clustering bases with different weighting of variables. *Psychometrika, 49*, 57-78.

DeSarbo, W. S., & Grisaffe, D. (1998). Combinatorial optimization approaches to constrained market segmentation: An application to industrial market segmentation. *Marketing Letters, 9*, 115-134.

Diday, E. (1986). Orders and overlapping clusters by pyramids. In J. de Leeuw, W. J. Heiser, J. Meulman, & F. Critchley (Eds.), *Multidimensional data analysis* (pp. 201-234). Leiden: DSWO Press.

Diehr, G. (1985). Evaluation of a branch and bound algorithm for clustering. *SIAM Journal for Scientific and Statistical Computing, 6*, 268-284.

Draper, N. R., & Smith, H. (1981). Applied regression analysis (2nd edition). New York: Wiley.

Durand, C., & Fichet, B. (1988). One-to-one correspondence in pyramidal representations: A unified approach. In H. H. Bock (Ed.), *Classification and related methods of data analysis* (pp. 85-90). New York: Springer-Verlag.

Ferligoj, A., & Batagelj, V. (1992). Direct multicriteria clustering algorithms. *Journal of Classification, 9*, 43-61.

Fisher, R. A. (1936). The use of multiple measurements in taxonomic problems. *Annals of Eugenics, 7*, 179-188.

Flueck, J. A., & Korsh, J. F. (1974). A branch search algorithm for maximum likelihood paired comparison ranking. *Biometrika, 61*, 621-626.

Forgy, E. W. (1965). Cluster analyses of multivariate data: Efficiency versus interpretability of classifications. *Biometrics, 21*, 768.

Fowlkes, E. B., Gnanadesikan, R., & Kettenring, J. R. (1988). Variable selection in clustering. *Journal of Classification, 5*, 205-228.

Fowlkes, E. B., & Mallows, C. L. (1983). A method for comparing two hierarchical clusterings. *Journal of the American Statistical Association, 78*, 553-584.
Friedman, J. H., & Meulman, J. J. (2004). Clustering objects on subsets of attributes. *Journal of the Royal Statistical Society B, 66*, 815-849.
Fukunaga, K. (1990). *Introduction to statistical pattern recognition* (2nd ed.). New York: Academic Press.
Furnival, G. M. (1971). All possible regressions with less computation. *Technometrics, 13*, 403-408.
Furnival, G. M., & Wilson, R. W. (1974). Regression by leaps and bounds. *Technometrics, 16*, 499-512.
Gnanadesikan, R., Kettenring, J. R., & Tsao, S. L. (1995). Weighting and selection of variables for cluster analysis. *Journal of Classification, 12*, 113-136.
Green, P. E., Carmone, F. J., & Kim, J. (1990). A preliminary study of optimal variable weighting in K-means clustering. *Journal of Classification, 7*, 271-285.
Groenen, P. J. F., & Heiser, W. J. (1996). The tunneling method for global optimization in multidimensional scaling. *Psychometrika, 61*, 529-550.
Groenen, P. J. F., Heiser, W. J., & Meulman, J. J. (1999). Global optimization of least-squares multidimensional scaling by distance smoothing. *Journal of Classification, 16*, 225-254.
Grötschel, M., Jünger, M., & Reinelt, G. (1984). A cutting plane algorithm for the linear ordering problem. *Operations Research, 32*, 1195-1220.
Guénoche, A. (1993). Enumération des partitions de diamètre minimum. *Discrete Mathematics, 111*, 277-287.
Guénoche, A. (2003). Partitions optimisées selon différents critères: Enumération des partitions de diamètre minimum. *Mathematics and Social Sciences, 41*, 41-58.
Guénoche, A., Hansen, P., & Jaumard, B. (1991). Efficient algorithms for divisive hierarchical clustering with the diameter criterion. *Journal of Classification, 8*, 5-30.
Hamamoto, Y., Uchimura, S., Matsura, Y., Kanaoka, T., & Tomita, S. (1990). Evaluation of the branch and bound algorithm for feature selection. *Pattern Recognition Letters, 11*, 453-456.
Hand, D. J. (1981a). Branch and bound in statistical data analysis. *The Statistician, 30*, 1-13.
Hand, D. J. (1981b). *Discrimination and classification.* New York: Wiley.
Hansen, P., & Delattre, M. (1978). Complete-link cluster analysis by graph coloring. *Journal of the American Statistical Association, 73*, 397-403.
Hansen, P., & Jaumard, B. (1987). Minimum sum of diameters clustering. *Journal of Classification, 2*, 277-291.
Hansen, P., & Jaumard, B. (1997). Cluster analysis and mathematical programming. *Mathematical Programming, 79*, 191-215.
Hansen, P., Jaumard, B., & Mladenovic, N. (1998). Minimum Sum of squares clustering in a low dimensional space. *Journal of Classification, 15*, 37-55.

Hartigan, J. A. (1975). *Clustering algorithms*. New York: Wiley.
Hastie, T., Tibshirani, R., & Friedman, J. (2001). *The elements of statistical learning*. Berlin: Springer-Verlag.
Heinz, G., Peterson, L. J., Johnson, R. W., & Kerk, C. J. (2003). Exploring relationships in body dimensions. *Journal of Statistics Education, 11*, www.amstat.org/publications/jse/v11n2/datasets.heinz.html.
Heiser, W. J. (1988). Selecting a stimulus set with prescribed structure from empirical confusion frequencies. *British Journal of Mathematical and Statistical Psychology, 41*, 37-51.
Held, M., & Karp, R. M. (1962). A dynamic programming approach to sequencing problems. *Journal of the Society for Industrial and Applied Mathematics, 10*, 196-210.
Hocking, R. R., & Leslie, R. N. (1967). Selection of the best subset in regression analysis. *Technometrics, 9*, 531-540.
Hosmer, D. W., Jovanovic, B., & Lemeshow, S. (1989). Best subsets logistic regression. *Biometrics, 45*, 1265-1270.
Hubert, L. (1973). Monotone invariant clustering procedures. *Psychometrika, 38*, 47-62.
Hubert, L. J. (1974). Some applications of graph theory to clustering. *Psychometrika, 39*, 283-309.
Hubert, L. (1976). Seriation using asymmetric proximity measures. *British Journal of Mathematical and Statistical Psychology, 29*, 32-52.
Hubert, L. J. (1987). *Assignment methods in combinatorial data analysis*. New York: Marcel Dekker.
Hubert, L., & Arabie, P. (1985). Comparing partitions. *Journal of Classification, 2*, 193-218.
Hubert, L. J., & Arabie, P. (1986). Unidimensional scaling and combinatorial optimization. In J. de Leeuw, W. Heiser, J. Meulman, and F. Critchley (Eds.), *Multidimensional data analysis* (pp. 181-196). Leiden, Netherlands: DSWO Press.
Hubert, L., & Arabie, P. (1994). The analysis of proximity matrices through sums of matrices having (anti-) Robinson forms. *British Journal of Mathematical and Statistical Psychology, 47*, 1-40.
Hubert, L., Arabie, P., & Meulman, J. (1998). Graph-theoretic representations for proximity matrices through strongly anti-Robinson or circular strongly anti-Robinson matrices. *Psychometrika, 63*, 341-358.
Hubert, L., Arabie, P., & Meulman, J. (2001). *Combinatorial data analysis: Optimization by dynamic programming*. Philadelphia: Society for Industrial and Applied Mathematics.
Hubert, L., Arabie, P., & Meulman, J. (2002). Linear unidimensional scaling in the L_2-norm: Basic optimization methods using MATLAB. *Journal of Classification, 19*, 303-328.
Hubert, L. J., & Baker, F. B. (1978). Applications of combinatorial programming to data analysis: The traveling salesman and related problems. *Psychometrika, 43*, 81-91.

Hubert, L. J., & Golledge, R. G. (1981). Matrix reorganization and dynamic programming: Applications to paired comparisons and unidimensional seriation. *Psychometrika, 46*, 429-441.

Hubert, L., & Schultz, J. (1975). Maximum likelihood paired comparison ranking and quadratic assignment. *Biometrika, 62*, 655-659.

Hubert, L. J., & Schultz, J. (1976). Quadratic assignment as a general data analysis strategy. *British Journal of Mathematical and Statistical Psychology, 29*, 190-241.

Johnson, E. L., Mehrotra, A., & Nemhauser, G. L. (1993). Min-cut clustering. *Mathematical Programming, 62*, 133-151.

Johnson, S. C. (1967). Hierarchical clustering schemes. *Psychometrika, 32*, 241-254.

Kendall, D. G. (1971a). Abundance matrices and seriation in archaeology. *Zeitschrift für Wahrscheinlichkeitstheorie, 17*, 104-112.

Kendall, D. G. (1971b). Seriation from abundance matrices. In F. R. Hodson, D. G. Kendall, & P. Tăutu (Eds.), *Mathematics in the archaeological and historical sciences* (pp. 215-252). Edinburgh: Edinburgh University Press.

King, J. E. (2003). Running a best subsets logistic regression: An alternative to stepwise methods. *Educational and Psychological Measurement, 63*, 392-403.

Kittler, J. (1978). Feature set search algorithms. In C. H. Chen (Ed.), *Pattern recognition and signal processing* (pp. 41-60). The Netherlands: Sijthoff and Noordhoff.

Klastorin, T. D. (1985). The p-median problem for cluster analysis: A comparative test using the mixture model approach. *Management Science, 31*, 84-95.

Klein, G., & Aronson, J. E. (1991). Optimal clustering: A model and method. *Naval Research Logistics, 38*, 447-461.

Koontz, W. L. G., Narendra, P. M., & Fukunaga, K. (1975). A branch and bound clustering algorithm. *IEEE Transaction on Computers, C-24*, 908-915.

Koopmans, T. C., & Beckmann, M. (1957). Assignment problems and the location of economic activities. *Econometrica, 25*, 53-76.

Krieger, A. M., & Green, P. E. (1996). Modifying cluster-based segments to enhance agreement with an exogenous response variable. *Journal of Marketing Research, 33*, 351-363.

LaMotte, L. R., & Hocking, R. R. (1970). Computational efficiency in the selection of regression variables. *Technometrics, 12*, 83-93.

Land, A. H., & Doig, A. G. (1960). An automatic method for solving discrete programming problems. *Econometrica, 28*, 497-520.

Lawler, E. L. (1964). A comment on minimum feedback arc sets. *IEEE Transactions on Circuit Theory, 11*, 296-297.

Leenen, I., & Van Mechelen, I. (1998). A branch-and-bound algorithm for Boolean regression. In I. Balderjahn, R. Mathar, & M. Schader (Eds.), *Data highways and information flooding: A challenge for classification and data analysis* (pp. 164-171). Berlin: Springer-Verlag.

Little, J. D. C., Murty, K. G., Sweeney, D. W., & Karel, C. (1963). An algorithm for the traveling salesman problem. *Operations Research, 11*, 972-989.

MacQueen, J. B. (1967). Some methods for classification and analysis of multivariate observations. In L. M. Le Cam & J. Neyman (Eds.), *Proceedings of the fifth Berkeley symposium on mathematical statistics and probability* (vol. 1, pp. 281-297). Berkeley, CA: University of California Press.

Mallows, C. L. (1973). Some comments on C_p. *Technometrics, 15*, 661-675.

Manning, S. K., & Shofner, E. (1991). Similarity ratings and confusability of lipread consonants compared with similarity ratings of auditory and orthographic stimuli. *American Journal of Psychology, 104*, 587-604.

MathWorks, Inc. (2002). *Using MATLAB (Version 6)*. Natick, MA: The MathWorks, Inc.

Miller, A. J. (2002). *Subset selection in regression* (2nd ed.). London: Chapman and Hall.

Miller, G. A., & Nicely, P. E. (1955). Analysis of perceptual confusions among some English consonants. *Journal of the Acoustical Society of America, 27*, 338-352.

Milligan, G. W. (1989). A validation study of a variable-weighting algorithm for cluster analysis. *Journal of Classification, 6*, 53-71.

Milligan, G. W. (1996). Clustering validation: Results and implications for applied analyses. In P. Arabie, L. J. Hubert, & G. De Soete (Eds.), *Clustering and classification* (pp. 341-375). River Edge, NJ: World Scientific Publishing.

Milligan, G. W., & Cooper, M. C. (1986). A study of the comparability of external criteria for hierarchical cluster analysis. *Multivariate Behavioral Research, 21*, 441-458.

Minitab, Inc. (1998). *Minitab user's guide 2: Data analysis and quality tools*. State College, PA: Minitab, Inc.

Morgan, B. J. T., Chambers, S. M., & Morton, J. (1973). Acoustic confusion of digits in memory and recognition. *Perception & Psychophysics, 14*, 375-383.

Mulvey, J., & Crowder, H. (1979). Cluster analysis: An application of Lagrangian relaxation. *Management Science, 25*, 329-340.

Murty, K. G. (1995). *Operations research: Deterministic optimization models*. Englewood Cliffs, NJ: Prentice-Hall.

Murty, K. G., Karel, C., & Little, J. D. C. (1962). *The traveling salesman problem: Solution by a method of ranking assignments*. Cleveland: Case Institute of Technology.

Narendra, P. M., & Fukunaga, K. (1977). A branch and bound algorithm for feature subset selection. *IEEE Transactions on Computers, 26*, 917-922.

Neter, J., Wasserman, W., & Kutner, M. H. (1985). *Applied linear statistical models* (2nd ed.). Homewood, IL: Irwin.

Palubeckis, G. (1997). A branch-and-bound approach using polyhedral results for a clustering problem. *INFORMS Journal on Computing, 9*, 30-42.

Parker, R. G., & Rardin R. L. (1988). *Discrete optimization*. San Diego: Academic Press.

Phillips, J. P. N. (1967). A procedure for determining Slater's *i* and all nearest adjoining orders. *British Journal of Mathematical and Statistical Psychology, 20*, 217-225.

Phillips, J. P. N. (1969). A further procedure for determining Slater's *i* and all nearest adjoining orders. *British Journal of Mathematical and Statistical Psychology, 22*, 97-101.

Ramnath, S., Khan, M. H., & Shams, Z. (2004). New approaches for sum-of-diameters clustering. In D. Banks, L. House, F. R. McMorris, P. Arabie, & W. Gaul (Eds.), *Classification, clustering, and data mining applications* (pp. 95-103). Berlin: Springer-Verlag.

Ranyard, R. H. (1976). An algorithm for maximum likelihood ranking and Slater's *i* from paired comparisons. *British Journal of Mathematical and Statistical Psychology, 29*, 242-248.

Rao, M. R. (1971). Cluster analysis and mathematical programming. *Journal of the American Statistical Association, 66*, 622-626.

Ridout, M. S. (1988). Algorithm AS233: An improved branch and bound algorithm for feature subset selection. *Applied Statistics, 37*, 139-147.

Roberts, S. J. (1984). Algorithm AS199: A branch and bound algorithm for determining the optimal feature subset of a given size. *Applied Statistics, 33*, 236-241.

Robinson, W. S. (1951). A method for chronologically ordering archaeological deposits. *American Antiquity, 16*, 293-301.

Rodgers, J. L., & Thompson, T. D. (1992). Seriation and multidimensional scaling: A data analysis approach to scaling asymmetric proximity matrices. *Applied Psychological Measurement, 16*, 105-117.

Ross, B. H., & Murphy, G. L. (1999). Food for thought: Cross-classification and category organization in a complex real-world domain. *Cognitive Psychology, 38*, 495-553.

Simantiraki, E. (1996). Unidimensional scaling: A linear programming approach minimizing absolute deviations. *Journal of Classification, 13*, 19-25.

Slater, P. (1961). Inconsistencies in a schedule of paired comparisons. *Biometrika, 48*, 303-312.

Soland, R. M. (1979). Multicriteria optimization: A general characterization of efficient solutions. *Decision Sciences, 10*, 26-38.

Späth, H. (1980). *Cluster analysis algorithms for data reduction and classification of objects*. New York: Wiley.

Steinley, D. (2003). Local optima in *K*-means clustering: What you don't know may hurt you. *Psychological Methods, 8*, 294-304.

Thurstone, L. L. (1927). The method of paired comparisons for social values. *Journal of Abnormal and Social Psychology, 31*, 384-400.

Tobler, W. R. (1976). Spatial interaction patterns. *Journal of Environmental Studies, 6*, 271-301.

van der Heijden, A. H. C., Mathas, M. S. M., & van den Roovaart, B. P. (1984). An empirical interletter confusion matrix for continuous-line capitals. *Perception & Psychophysics, 35*, 85-88.

Vega-Bermudez, F., Johnson, K. O., & Hsiao, S. S. (1991). Human tactile pattern recognition: Active versus passive touch, velocity effects, and patterns of confusion. *Journal of Neurophysiology, 65*, 531-546.

Ward, J. H. (1963). Hierarchical grouping to optimize an objective function. *Journal of the American Statistical Association, 58*, 236-244.

Weisberg, S. (1985). *Applied linear regression.* New York: Wiley.

Zupan, J. (1982). *Clustering of large data sets.* Letchworth, UK: Research Studies Press.

Index

adjacency test, 100, 116, 132
adjusted Rand index, 177
agglomerative hierarchical clustering, 15
anti-Robinson patterning, 113-115

bipartitioning, 15, 25
Boolean regression, 199
bounding, 6
bound test, 100, 121, 136
branching, 6
branch forward, 6, 21-22
branch right, 6, 21-22

cluster analysis, 9, 15-87
   hierarchical methods, 15
   partitioning methods, 15-87
coefficient of determination, 188
coloring (of threshold graphs), 25-26
combinatorial data analysis, vii, 1
complete-linkage clustering, 15, 25-27
compact partitioning, 25
complete solution, vii
computer programs
   bbdiam.for, 39
   bbdisum.for, 40
   bbdom.for, 111
   bbforwrd.for, 145
   bbinward.for, 144
   bbbiwcss.for, 85
   bburcg.for, 127
   bburg.for, 127
   bbwcss.for, 73
   bbwcsum.for, 56
   bbwrcg.for, 127
   bbwrg.for, 127
   bestsub.for, 199
   bestsub.m, 199

Defays' maximization (criterion), 129-130
dependent variable (regression), 187
diameter, 17
  cluster, 17

maximum method, 25
partition, 17
reduction ratio, 35
discrete optimization problems, 1
dispensation, 21-22
divisive hierarchical clustering, 15
dominance index, 97-112
   adjacency test, 100
   available software, 111-112
   bound test, 100-102
   demonstration, 102-103
   description, 97
   extracting and ordering subsets, 103-110
   fathoming tests, 98-102
   interchange test, 107-108
   perfect dominance, 104
   strengths and limitations, 110
   tournament matrix, 103-104
dynamic programming, viii
   for seriation, 92

efficient frontier, 149
efficient set, 149
exchange algorithm / heuristic, 27-28
   pairwise interchange, 27
   single-object relocation 27

fathoming (test), 94
feasible completion test, 21-22
feature selection (see variable selection)
forward selection (regression), 194

gradient indices, 113-128
   adjacency test, 116-120
   anti-Robinson patterning, 113-115
   available software, 127-128
   bound test, 121-123
   demonstration, 123-126
   fathoming tests, 115-123
   strengths and limitations, 126
   unweighted within row, 114
   unweighted within row / column, 114

weighted within row, 115
weighted within row / column, 115

independent variable (regression), 187
initial upper (lower) bound, 7
interchange test, 132-136

K-means, 59-61
knapsack problem, 5

lipread consonants data, 34-37, 53-54
logistic regression, 199

Mallows C_p, 200
masking variables
MATLAB®, 199
minimum backtracking row layout
    problem, 2
minimum-diameter partitioning, 25-41
  available software, 39-41
  initialize step, 26-30
  overview, 25-26
  numerical examples, 32-37
  partial solution evaluation, 30-32
  strengths and limitations, 39
  sum of cluster diameters, 38
monotonicity (property), 174
multicollinearity, 198
multiobjective cluster analysis, 77
multiobjective partitioning, 77-87
  available software, 85-87
  biobjective partitioning, 81-82
  multiple bases, 77-82
  multiple criteria, 82-84
  strengths and limitations, 84-85
multiobjective seriation, 147-169
  description, 147-149
  dominance index (multiple matrices),
    149-153
  efficient (nondominated) solutions, 149
  multiple gradient indices, 160-164
  multiple matrices and multiple criteria,
    164-168
  normalized weights, 148
  strengths and limitations, 169
  unidimensional scaling (multiple
    matrices), 153-159

nondominated (solution), 82, 149

optimal (solution), vii

p-median problem, 5
paired-comparison matrix, 103, 108
partial sequence, 6
Pareto (see efficient set), 149
partial enumeration strategies, 3
partial solution, vii
partial solution evaluation, 21-22
partitioning, 15-87
  branch-and-bound algorithm, 20-24
  indices 16-19
  number of feasible partitions, 18-19
perfect dominance, 104
pruning, 6

quadratic assignment problem, 5

R^2 (see coefficient of determination)
redundancy test, 93-94
regression, 187-202
residual sum of squares, 188
retracting (retraction), 6, 8, 21, 23
row and column gradients, 113-114
row gradients, 113-114

scalar-multiobjectiive-programming-
    problem (SMPP), 147
seriation, 91-169
  applications, 91
  branch-and-bound paradigm, 93-95
  dynamic programming, 92
set partitioning problem, 5
slope coefficient, 187
sum of cluster diameters, 38
symmetric dissimilarity matrix, 15
symmetry test, 131

tournament matrix, 103-104
traveling salesman problem, 5
true (clustering) variables, 177

UDS (see unidimensional scaling)
unidimensional scaling, 129-146
  adjacency test, 132
  available software, 144-146
  bound test, 136
  demonstrations, 139-143
  description, 129-130
  fathoming tests, 131-139

lower bound, 131
  strengths and limitations, 144
  symmetry test, 131

variable selection, 9, 173-202
  in cluster analysis, 177-185
    branch-and-bound algorithm, 178-181
    demonstration, 182-183
    masking variables, 177
    strengths and limitations, 183-185
    true variables, 177
  in regression, 187-202
    available software, 199-202
    body measurements data, 193-198
    numerical example, 190-193
    regression analysis overview, 187-190
    strengths and limitations, 198-199
variable weighting, 177

Ward's minimum variance method, 15
within-cluster sum of squares partitioning 18, 59-76
  available software, 73-76
  initialize step, 60-63
  numerical examples, 66-71
  partial solution evaluation, 64-66
  relevance, 59-60
  strengths and limitations, 71-72
within-cluster sums of dissimilarities partitioning, 18, 43-58
  available software, 56-58
  initialize step, 44-46
  numerical examples, 50-54
  overview, 43
  partial solution evaluation, 46-50
  strengths and limitations, 54-56

springeronline.com

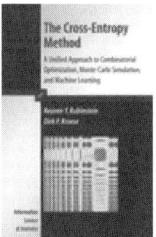

The Cross-Entropy Method

R.Y. Rubinstein and D.P. Kroese

The cross-entropy (CE) method is one of the most significant developments in randomized optimization and simulation in recent years. This book explains in detail how and why the CE method works. The CE method involves an iterative procedure where each iteration can be broken down into two phases: (a) generate a random data sample (trajectories, vectors, etc.) according to a specified mechanism; (b) update the parameters of the random mechanism based on this data in order to produce a ``better" sample in the next iteration. The simplicity and versatility of the method is illustrated via a diverse collection of optimization and estimation problems. The book is aimed at a broad audience of engineers, computer scientists, mathematicians, statisticians and in general anyone, theorist and practitioner, who is interested in smart simulation, fast optimization, learning algorithms, and image processing.

2004. 300 p. (Information Science and Statistics) Hardcover
ISBN 0-387-21240-X

An Introduction to Rare Event Simulation

J.A. Bucklew

This book presents a unified theory of rare event simulation and the variance reduction technique known as importance sampling from the point of view of the probabilistic theory of large deviations. This perspective allows us to view a vast assortment of simulation problems from a unified single perspective. This text keeps the mathematical preliminaries to a minimum with the only prerequisite being a single large deviation theory result that is given and proved in the text. It concentrates on demonstrating the methodology and the principal ideas in a fairly simple setting. It includes detailed simulation case studies covering a wide variety of application areas including statistics, telecommunications, and queueing systems.

2004. 260 p. (Springer Series in Statistics) Hardcover ISBN 0-387-20078-9

Easy Ways to Order ▶ Call: Toll-Free 1-800-SPRINGER • E-mail: orders-ny@springer.sbm.com • Write: Springer, Dept. S8113, PO Box 2485, Secaucus, NJ 07096-2485 • Visit: Your local scientific bookstore or urge your librarian to order.